OILGATE

'No other country does more in the field of
sanctions than the United Kingdom. I do not
claim any special virtue for this. It is Britain's
duty to do it.'
George Thomson, Commonwealth Secretary,
March 1968

'No company in which we have an interest is
supplying to Rhodesia.'
Sir Frank McFadzean, Chairman of Shell,
June 1976

' . . . neither Shell and BP nor any company in
which they have an interest have engaged, either
directly or indirectly, in supplying crude oil
or oil products to Rhodesia.'
Letter from British Government to UN,
September 1976

'I am generally satisfied with the observance
of sanctions by United Kingdom firms, but
where we have evidence suggesting breaches
of sanctions we certainly investigate these.
Our record in this respect is second to none.'
Ted Rowlands, Minister of State at the Foreign
Office, March 1977

OILGATE
The Sanctions Scandal

Martin Bailey

CORONET BOOKS
Hodder and Stoughton

Printed and bound in Great Britain for
Hodder and Stoughton Paperbacks, a
division of Hodder and Stoughton Ltd.,
Mill Road, Dunton Green, Sevenoaks,
Kent (Editorial Office: 47 Bedford
Square, London, WC1 3DP) by
Hunt Barnard Printing Ltd.,
Aylesbury, Bucks.

ISBN 0 340 24488 7

OILGATE Contents

ACKNOWLEDGEMENTS

First I must thank the 'deep throats' within the oil companies who were willing to talk about the sensitive issue of sanctions-busting. In some cases they even took the risks involved in providing documentary evidence. Without their help it is fair to say, this book would probably never have come to be written. Obviously, these people still have to remain unnamed. I have given some details of the role they played, in order to help explain how the sanctions story was broken, but I have also been very careful not to divulge any information which might enable the oil companies to trace the identity of our informants.

In the case of a story one has followed for many years, there are inevitably so many people who have helped in so many different ways. But I would like to single out the following in particular: Sietse Bosgra, Alan Gordon Walker, David Haslam, Cor Groenendijk, Jorge Jardim, Pat Kavanagh, Peter Kellner, Linda Lennard, Bruce Page, Andrew Phillips, Alun Roberts, Anthony Sampson, Graca da Silva, and Gloria Young. I would also like to thank the Haslemere Group and the American Committee on Africa which raised funds from anti-apartheid organisations to help subsidise our research. My special thanks are due to Paul Manning, who skilfully edited my manuscript.

But there is one person, above all, who has played a very special part in this book. Bernard Rivers has been my colleague on the sanctions trail for several years. Bernard played a key role in exposing how Mobil was supplying Rhodesia through a clandestine paper-chase. Perhaps the best way I can thank Bernard for his part in the story is by writing this book.

INTRODUCTION

by Anthony Sampson

The importance of this book lies not just in the story it tells, remarkable though that is, but in the sharp and sudden light it throws on the intimate workings of government, diplomacy and business. The word 'scandal' often implies a frivolous or muck-raking interest; but in fact any major political scandal has a much more serious content, for historians and political scientists as well as for journalists and the general public. For it is only when such scandals are pursued and exposed that the dark skies of government and politics are suddenly lit up as if by lightning, revealing not only the wide landscape, but the details and postures which can explain so much about how business is actually done. Just as suddenly the light disappears and the scene is once again enclosed in darkness.

The serious study of this kind of scandal can provide a rare case-history of how politics really operate, which no one in the field can afford to ignore. It was this that the Watergate scandal in Washington provided; not only an extraordinary epic, but unique documentation about the workings of the White House. The title of this book, *Oilgate*, implies a resemblance to Watergate, and with good reason; for not only do the two stories have some important resemblances, including the network of lies, the political corruption, and the creeping cover-up and self-deception; but this story also provides a wealth of documentation and detail which makes everything it touches look different. Nor is it just a story about oil and Africa; it is about the whole character of government, the relations between politicians and civil servants, and between governments and multi-national corporations. In all these fields it provides new insights, backed up by authentic detail.

Its most obvious importance is in the context of Southern

Africa; and here what emerges most clearly is the extent of the institutionalised hypocrisy which has governed so much of the policy of Britain and the United States. Throughout this book Martin Bailey traces the contrasts between the public statements of the British Government, announcing its determination to bring down Ian Smith's rebel regime, and their actual connivance with the breaking of oil sanctions which was having the opposite effect. By putting together the historical jigsaw he shows, with devastating effect, how casual and matter-of-fact that hypocrisy and contradiction had become; how Ministers carefully refrained from asking the questions which would produce embarrassing answers, and how the whole Cabinet had become accustomed to blind eyes. The Government's excuse for this sustained humbug is fairly presented: that Britain could not afford to antagonise South Africa or risk losing her trade. And certainly there is plenty of evidence that the British voters, together with much of the press, were quite happy to be deceived, in order not to be faced with a painful decision. But what is most worrying to anyone concerned with democracy is how little chance the public were given, by any of the politicians concerned, to learn the true nature of that choice. The Government has assumed from the beginning that the public would simply prefer not to know.

Of course this anxiety to evade difficult choices has been implicit in much of the policy of Whitehall, and also of Washington, towards the mounting conflict between black and white Africa, where there are powerful economic and diplomatic interests on both sides. But the Rhodesian oil sanctions story points up the true cost of that evasion with terrifying detail. It was not simply the financial cost, in terms of extravagant contradictory operations, with the naval blockade of Beira proceeding while the oil companies were quietly permitted to transport oil across the land frontiers. Much more serious was the cost in terms of human suffering and bloodshed; for the failure to implement sanctions allowed the Smith regime to entrench itself, the results being, a decade later, a mounting civil war with nearly a thousand Rhodesians killed every month. That consequence could have been foreseen, and was foreseen by many critics of the Government. But the elaborate evasions continued without any such foresight, and one of the most alarming

8

features of the story is how few of the characters involved in the deception felt any need to protest or to set themselves against it. The clear and depressing message for Africa is that Western Governments, when faced with the need for a clear choice, will do everything they can to avoid it.

But this narrative tells us a great deal about the nature of government, particularly British Government, which goes beyond specifically African problems. For in these pages we have a new picture of how decisions can be taken in Whitehall without Parliament or the public being involved or consulted; and there are some sharp glimpses into the ways by which civil servants are able both to dominate their Ministers, and to collaborate with the executives of multinational corporations. We can observe how the oil companies were able to keep contact with the civil servants on a 'personal' basis – which meant, in effect, that the Ministers were not informed; how civil servants were dependent on the oil companies for their information, and subservient to their viewpoint; and how some influential civil servants later joined the oil industry. With this kind of symbiosis between Whitehall and the companies, the politicians rarely obtained any independent viewpoint, and politicians in the Ministry of Power, the department most involved with the oil embargo, continued in ignorance of the real relationships. Dr. Jeremy Bray, former Parliamentary Secretary to the Minister (Richard Marsh), later candidly admitted that until the story came to light a decade later he was 'not even aware that the Ministry of Power was the primary channel of communication with the oil companies over the Rhodesian embargo' (see p. 275).

Nearly all the civil servants and diplomats appeared eager to join the connivance and cover-up, but they also had a good deal of justification for assuming that the politicians did not wish to know too much; and when the Ministers did know, they were adept at evading the responsibility. The author rightly gives careful attention to the role of George Thomson (now Lord Thomson), the Commonwealth Minister who was responsible for approving the 'swap' by which the British oil companies sold oil to the French company, Total, which in turn supplied it to the Rhodesians; and perhaps the most significant revelations in this book are the descriptions of how the Government and the oil companies agreed on how to handle awkward Parliamentary questions about oil getting through to Rhodesia, <u>with answers which were clearly misleading</u>; and how soon afterwards these replies were foisted onto members of Parliament, providing (as the author says) '<u>blatant disinformation</u>'. It was a special irony that the Government were more worried about offending South Africa by revealing how British companies were swapping supplies, than with misleading their own Parliament.

The insights into Whitehall and Westminster are valuable to anyone who wants to understand how British Government works, for they provide the kind of poignant detail of the relationships from both sides that is not available in textbooks, in Prime Ministers' memoirs, or even in Richard Crossman's diaries. But the most important and unique parts of this book, I believe, are those which illustrate the characteristics and attitudes of the oil companies. For the whole study of the operations of <u>multinational corporations</u> is still in its infancy, and is afflicted by theories and generalisations which often have little relevance to actual situations. Here, by contrast, we see how individual executives inside Britain's two largest companies reacted to a specific situation; we are able to read their letters and minutes, and the records of their meetings – sometimes from more than one source; and we are able to visualise the predicaments of the executives, their Directors and Chairmen, as the story unfolds. In the process, many readers may gain a different impression of the nature of these remarkable industrial organisms. The giant companies often present themselves to the public and their shareholders as if they were a single unified organisation, as solid and structured as the skyscrapers that contain their headquarters; and their investment plans, and their building of massive pipelines and refineries, corroborate this view. But in many parts of this story the companies appear in a very different perspective; as if they were no more than <u>loose federations of subsidiaries</u> which are intensely vulnerable to pressures from local governments, and who find themselves risking their world reputation – and their supplies of oil from black Africa – for the sake of maintaining sales to a relatively tiny part of their total market.

The reluctance of multinational corporations to take sides has been a familiar phenomenon; before and during the

10

Second World War many of them, including Esso and ITT, maintained relations with the Nazis as well as the Allies. The oil companies can, and do, claim that their multinational character prevents them from showing a national bias, and that this can often be of great benefit: they point to their achievement in 1973, when they defied not only the Arab embargo but also the British Prime Minister Ted Heath by insisting that their oil should be equally shared among their client countries. But the problem of the ultimate loyalties and commitment of the global corporations when faced with a situation which calls for moral commitment remains very worrying, and the conflict in Southern Africa presents a kind of testing-ground; for the companies' insistence on maintaining their trade, of being all things to all men, can easily degenerate – as we can see in this book – into a determination to frustrate the policies of their home government, to the point of gross deception. For a country like South Africa which is threatened with sanctions and international isolation, these companies can prove the most valuable allies. As the Johannesburg *Financial Mail* commented in 1973 (when Iran and the oil companies were both protecting South Africa from the effects of the Arab embargo): 'There can be no greater blessing for South Africa – apart from the fact that Iran is well disposed – than that the oil business is still largely in the hands of international companies with no discernible leanings of excessive patriotism' (see p. 239). It was an equal blessing for Rhodesia; and for South Africa today – with Iranian supplies cut off, but with oil still mysteriously arriving – the blessing conferred by these many-sided companies is still more apparent.

Many other characteristics of the big oil companies are evident in the course of this narrative. Through the company memoranda and records we can almost see the winks and nods and careful silences by which the critical questions about sanctions-breaking are evaded, with the use of such phrases as 'grey areas', 'Friends to the North' and 'extraneous demand' to describe Rhodesia, or the word 'cosmetic' to mean, in effect, mendacious. The readiness of the companies to mislead both the public and the Government sometimes seems almost congenital. Perhaps the most remarkable instance occurred as late as September 1978, when the companies had already been shown to have pro-

11

duced a torrent of misleading information. The author, together with two colleagues, Bernard Rivers and Peter Kellner, published a well-informed article in the London *Sunday Times*, alleging that the oil companies were *still* enabling oil to get through to Rhodesia through a swap arrangement with the South African company, Sasol. BP reacted immediately with a highly misleading denial and then (on the principle apparently of 'shoot first, ask questions afterwards') sent out an investigating team to South Africa, where they received an assurance from the local Chairman that the swap had been terminated. The Chairman of BP, Sir David Steel, then made a further statement in London denying any continuing swap, and describing the *Sunday Times* article as both 'offensive and defamatory'; and their denial was now backed up by Shell. Only at this point did BP in South Africa, when an employee checked the computer print-outs, discover that the swap was in fact still continuing and take steps finally to end it; and after sending out another investigating team, BP at last admitted to the *Sunday Times* that their previous denial, as they put it, 'could have given a misleading impression of the situation'. This experience reveals all too clearly the immense difficulty of persuading oil companies to either discover or reveal the truth.

It is part of the justification for the evasiveness of the oil companies that they must remain politically neutral in order to retain their global operations under so many different regimes; but the revelations in this story cast some doubt on this neutrality – at least in the case of Shell. The then Chairman of Shell, Sir Frank McFadzean, claimed that he laid special emphasis on local companies avoiding involvement in national politics; but as the author points out, his own political attitudes were sharply defined. McFadzean, it turns out, had already by 1966 established close personal relations with the South African Prime Minister, John Vorster; and in February 1968, at his meeting with George Thomson, he explained that he was worried that the British refusal to sell arms to South Africa was leading to deteriorating relations between the two countries (see p. 264). It might seem remarkable that the head of a global company, with huge interests in Nigeria as well as many Arab states, should have become so actively involved in supporting South

12

Africa; and it might well upset many shareholders to discover that their company was, in effect, part of the South African arms lobby.

The evidence presented in this book will certainly provide much new material for the continuing debate about the power of multinational corporations, and the problems of controlling them; and even Harold Wilson, who has until recently been reluctant to criticise these companies, has admitted that the evidence about oil sanctions suggests that the attacks have been understated, rather than overstated, 'certainly in respect of Shell, arrogantly asserting power with scant regard for responsibility' (see p. 266). Of course it is very possible for Shell to reply, as they have, that the Labour Government was only too glad to evade the issues, and that they used the companies for that purpose. But the notion that the companies were merely passive instruments is effectively contradicted in these pages; and it becomes clear that Shell, at least, had a fixed determination to support South Africa and to allow oil to Rhodesia; and was prepared to go to considerable lengths to execute it.

In view of the record of evasiveness and mendacity from both companies and Governments, the most remarkable fact about this story is that it ever came to light. This book has a double historical interest. It not only provides a careful account of the events; but the author himself played an important part in the exposure of the scandal; without the persistence and patience of Martin Bailey and his colleague Bernard Rivers it is very doubtful whether the true facts would ever have emerged. The account in the first chapters of how Bailey and Rivers were able gradually to uncover the story, with the help of sources inside the companies and leaked documents, is itself an important piece of history; and it glaringly reveals the inadequacy of existing machinery in Britain for investigation in this kind of critical area, whether by the Government or by Parliament, or by the Press. The newspapers, after their first excitement about the failure of sanctions and the Beira Patrol, lost interest in the subject and it was left to these two freelance journalists to pursue it over a period of four years; and even when the evidence was finally assembled, it was difficult to interest the media. It is therefore especially appropriate that Bailey and Rivers, together with Peter Kellner of the *Sunday Times*, should

13

recently have received the British Press Award as Journalists of the Year.

It was the evidence revealed by Martin Bailey and Bernard Rivers which compelled the Foreign Secretary, Dr. David Owen, to appoint the commission headed by Thomas Bingham QC, which led to the publication of the Bingham Report in 1978. The Report itself was an historic document, providing new and incontrovertible evidence of the details of sanctions-breaking and the conspiracy of the cover-up; but it is a long and detailed document which is not really accessible to the general reader. This book therefore makes available much of the material for the first time to the public, against the wider background of the political crisis, and the author's own extensive inside information.

Oilgate then, is part of history; but it is not just past history. For it illuminates the character of institutions, and the problems of the democratic control of them, which will be with us for the foreseeable future; and the central problem of enforcing sanctions is one that is likely to become increasingly urgent, as the position in Southern Africa worsens. Many people, particularly those inside the big corporations, will doubtless maintain that the evidence in this book shows that sanctions are not, and may never be, workable. Certainly it shows the vast gap that exists between legislation and execution; and what emerges most clearly of all is the lack of the political will to insist on enforcement and to overcome the deliberate obstacles. But as the bloodshed in Southern Africa continues, the alternatives to sanctions appear all the more terrible, and the cost of the evasions and indecisions of the West become more evident. The true lesson of this book is surely that the effective implementation of any serious policy, including sanctions, depends on honest government and honest corporations, both of which have been so tragically found wanting.

14

AUTHOR'S FOREWORD

Soon after UDI Wilson said that Rhodesia was 'the greatest moral issue which Britain has had to face in the post-war world'.[1] But the British Government never lived up to the challenge. Sanctions, Britain's only weapon, failed to restore legality, and as a result a tiny white community successfully defied the entire international community for over thirteen years.

It is astonishing that the Oilgate story took so many years to break. Everyone always knew that Rhodesia was importing oil – how otherwise could the country continue to run? But what was *not* known, at least until recently, was the fact that it was the oil companies themselves which were sending in the oil. As Liberal leader David Steel explained:

> Those who shipped in the oil were not hostile powers. They were British companies, backed by the British Foreign Office, with the connivance of British Cabinet Ministers and the knowledge of the Prime Minister . . . The sanctions-busters were our own leaders![2]

The British Government connived with the country's two largest companies for over a decade to cover up a deception of quite staggering proportions.

The costs of UDI have been enormous. For the African population of Rhodesia it has meant another thirteen years of minority rule. At one time, sanctions offered an initial hope of a peaceful solution. But when they proved ineffective – and this became clear 'within weeks rather than months' – the Rhodesian nationalists had only one weapon left: the gun. The British Government and the international oil companies must therefore bear much of the responsibility for the suffering which has been caused by the escalating guerrilla war in Zimbabwe – Rhodesia.

For Britain the costs of UDI have been high, both financially and in terms of the country's battered international reputation. The Beira Patrol, however, has come to symbolise the costs of the Rhodesian débâcle. While the Royal Navy was proudly patrolling the port of Beira, refined oil was continuing to flow several hundred miles to the south through Lourenço Marques. The Beira Patrol probably cost the British taxpayer over £100 million. As one embittered letter-writer to The *Guardian* suggested: 'Presumably the Government calculated that the profits from the sale of oil to Rhodesia would offset the cost of the naval blockade that was supposed to stop it getting there.'[3]

This book tells the story of how the Rhodesians survived the United Nations oil embargo. But the definitive account of the role of the oil companies and the British Government in defying the sanctions the British Government itself had imposed, remains to be written. There are still a number of areas where considerable further research still needs to be done – particularly on the part played by the civil servants and the politicians.

In February 1979 the Labour Government proposed the establishment of a Parliamentary Special Commission on Oil Sanctions to determine whether Parliament or Ministers were misled over the oil embargo. After the proposed inquiry had been rejected by the House of Lords, it was still possible for the House of Commons to undertake the investigation by itself. But the Commission has still not yet been established.

Further efforts to cover up the sanctions scandal could well succeed. Only an inquiry armed with the powers to demand papers and require witnesses to testify is likely to get to the bottom of the matter – and answer the question of exactly where things went wrong. The implications of this scandal go way beyond Rhodesia: the sanctions story raises crucial questions about the way our system of government really works.

This book is divided into two parts. Part One – *Breaking the Story* – reveals how the sanctions cover-up was finally exposed. For a decade after the oil embargo was introduced, virtually nothing was published on the clandestine methods used by the Rhodesian regime to import their oil. The first part of this book therefore provides a chronological account from 1974, and explains how the different pieces of the Oil-

16

gate jig-saw puzzle were gradually assembled.

Part Two – *Busting Sanctions* – tells the story of how the oil companies conspired to supply Rhodesia with oil. It too is arranged chronologically, from 1965 to the present. The role of Shell and BP is stressed, largely because more information has emerged on the British oil companies; but as the book makes clear, Mobil, Caltex, and Total were just as deeply involved.

Extracts from the Bingham Report are quoted extensively. References to the report are given in the text by paragraph number – *e.g.* (B2.22). References to the documents contained in Appendix II of the Bingham Report refer to the page numbers of the edition published by Her Majesty's Stationary Office – *e.g.* (B/A222). It should be noted, however, that the page numbers of the original typed edition of the Bingham Report are different from the HMSO edition. References to sources other than the Bingham Report are numbered and given at the end of each chapter.

Over the period covered by the sanctions story many of the civil servants and politicians concerned have been honoured with various titles. In general, however, I have referred to people by the name that they were known by when they played their most important role in the saga. (Sir Harold) Wilson and (Lord) Thomson, for instance, are therefore referred to simply by their surnames. My apologies for this sacrifice of protocol to clarity. Similarly, the Mozambican capital of Maputo is referred to by the name it held during the period when it was used as a transit port for Rhodesian oil – Lourenço Marques.

The expression 'international' oil companies is used to refer to the five oil companies which operate in Southern Africa – that is, Shell, BP, Mobil, Caltex, and Total. Sometimes the local subsidiaries, rather than their head offices, are referred to, but this should be clear from the context.

Finally, quantities of oil are given in barrels (one barrel consists of 35 gallons, and 8 barrels amount to 1 ton of oil).

Martin Bailey, May 1979

NOTES

1 *Daily Telegraph*, 28 August 1978.
2 *News of the World*, 10 September 1978.
3 *Guardian*, 12 September 1978.

Part One
BREAKING THE STORY

1 *The Oil Conspiracy*

September 1974 was a fascinating time to be in the Mozambican capital of Lourenço Marques. A military coup in Lisbon on 25 April had suddenly ended over forty years of fascist rule in Portugal. The coup also had tremendous implications for the future of Portugal's colonial empire in Africa. The new Government decided to negotiate with the FRELIMO guerrilla fighters who had already seized control of much of the rural areas of Mozambique; the dreaded Portuguese secret police had retreated into hiding; there was suddenly a new sense of freedom in Lourenço Marques. And for the first time it was possible to ask openly about Rhodesian sanctions-busting.

My colleague, Bernard Rivers, flew into Lourenço Marques on 16 September, just four days before the new Transitional Government of Mozambique brought FRELIMO to power. Bernard, who was then twenty-seven, was a freelance researcher on Third World affairs. He had been sent out to Lourenço Marques (shortly to be renamed Maputo) by Granada Television as a consultant for a World in Action film about the dramatic changes taking place in Mozambique.

Bernard quickly grasped the new opportunities that had suddenly opened up to investigate the failure of Rhodesian sanctions over the nine years since UDI. In his spare time he decided to try to unravel the secrets of the sanctions-busters who were trading with Rhodesia from Lourenço Marques. His decision marked the beginning of a trail that was to lead to some of the world's largest multinational companies, and to the British Government itself – for both, it emerged, had been party to what was later to be hailed as one of the major scandals of recent years. Needless to say, back in September 1974, Bernard could have had no idea what his initial inquiries would eventually uncover.

After Ian Smith's Unilateral Declaration of Independence in 1965, Britain had gone to the United Nations to call for sanctions to end the rebellion. Rhodesia was landlocked, and therefore completely dependent on its neighbours for access to the outside world. But two countries refused to accept the UN decision to impose sanctions. These were South Africa and Portugal (the ruler of Mozambique). Both South Africa and Mozambique bordered Rhodesia, and their support enabled the rebel regime to continue clandestine trade with the outside world. In 1974 there was no direct railway between Rhodesia and South Africa (although a line was under construction), and this meant that the route through Mozambique was particularly important. Even after sanctions, the Rhodesians continued to send their foreign trade up and down the railway line running to the Mozambique port of Lourenço Marques.

It seemed incredible to Bernard that after nearly a decade of oil sanctions, the details of this trade had never been exposed. Admittedly, in the first few months of sanctions, immediately after UDI, a number of newspaper reports had revealed some of the methods that the rebel regime was using to continue its foreign trade. But the international community quickly lost interest in sanctions, and after 1966 little was publicly known about the secret procedures which the Rhodesians used to evade UN sanctions. Bernard realised that if the methods used by the Rhodesians could be exposed, then action might be taken to tighten sanctions. And if sanctions could be made more effective, even at this late stage, this might help to force the rebel leader to step down and concede majority rule. The only alternative seemed to be an escalation of the guerrilla war which was already claiming a mounting number of lives.

Bernard began by asking himself what he thought were fairly obvious questions about sanctions. Plainly, Rhodesia could do without luxuries such as Scotch whisky or Swiss watches, and if necessary the country could also survive without more basic imported goods such as shirts or shoes. But oil was clearly in a special category. Rhodesia had no oil deposits of its own; its stockpile would have lasted only a matter of weeks; and every day the country consumed some 15,000 barrels of oil. Simple arithmetic suggested that it would take around 70 railway tank-cars to carry this quantity of fuel. Several train-loads of oil were therefore needed every day to sustain Rhodesia. A daily traffic of such volume would be difficult to disguise.

Simply by listening to bar gossip in Mozambique one could easily gather that most of Rhodesia's oil was sent up the railway line from Lourenço Marques. But the big secret was who was sending the fuel. Perhaps the arrangements were so complicated that they would prove impossible to unravel. Or maybe the system was altered so frequently that it would be difficult to pin down how it actually worked. Either way Bernard appreciated that his chances of unearthing the identity of Rhodesia's secret oil suppliers were low – but he could not resist the challenge.

Bernard visited the port, and photographed the tanker *Mobil Durban* unloading its cargo of refined oil – much of which, he suspected, was destined for Rhodesia. But he quickly realised that it was the next link in the chain which might be the more revealing: the rail link from Lourenço Marques to Rhodesia. And so on 22 September 1974, only six days after arriving in Mozambique, he went down to the marshalling yards on the outskirts of town to take a closer look at the long lines of railway tank-cars which transported oil.

On the tank-cars belonging to Rhodesia Railways the 'RR' which had originally been displayed on the sides in large letters had long since been painted over. But one piece of evidence had proved more difficult to hide. On the bogies, just above the wheels, the letters 'RR' were embossed. A clumsy attempt had been made to disguise these marks by sticking a small lump of tar over the incriminating letters.

Bernard then looked to see whether destination labels had been affixed to the tank-cars, and found that some of the tank-cars carried cards in a little box on the side, specifying the quantity and type of fuel which had been loaded. Some also bore the word *completo* – the Portuguese for 'full' – but at the time Bernard did not appreciate its significance. Later he was to discover that *completo* was used as a code-word meaning that the tank-car was to be despatched to Rhodesia. Not surprisingly, however, the labels gave no clue as to the ownership of the oil in the tank-cars, let alone the destination. Clearly careful measures had been taken to deter prying eyes. Bernard took one of the labels as evidence, and quickly moved away from the siding.

According to railway workers at the marshalling yard three train loads of tank-cars were despatched every day up the line to Rhodesia, normally leaving during the night – at 22.16, 0.50, and 3.00. The official explanation was that it was 'dangerous' to transport fuel in the heat of the African sun, but the real reason, an oil company employee later admitted, was to make this sensitive trade 'less visible'.

Bernard then found a contact in the statistics division of the Mozambique Railways who nervously revealed the quantities of oil sent up the line. Altogether a total of 4,750,000 barrels of oil products had been railed from Lourenço Marques into Rhodesia during the previous year. The size of this figure confirmed that virtually all of Rhodesia's oil imports passed through the Mozambican capital. But what it did not show was who had actually consigned the oil. After close questioning, the railway official reluctantly admitted that he had once seen a confidential memorandum which stated that most of the oil was sent by one particular firm. But he refused to name the company.

Was it one of the international oil companies who operated in Mozambique? No, the contact replied, it was actually a well-known shipping and forwarding firm. The furthest he would go was to state that it had 'an English-sounding name'. But what was this Company X? Clearly Mozambique Railways must be billing *someone* for sending the tank-cars up the line to Rhodesia. So Bernard returned to the railway office and eventually found a friendly clerk in the accounts department who provided the missing detail: Company X was a South African-based firm called Freight Services Limited, a company which was later revealed to have played a key role in the sanctions story.

The next stop was Salisbury. Much as he would have liked to have travelled on the railway that actually carried the oil, Bernard was told that the line did not take passengers, and instead he had to fly into the Rhodesian capital. Again he went first to the railway marshalling yard. The depot was no

22

place to linger in, since snoopers risked a heavy prison sentence, but from a brief inspection it was clear that the tank-cars from which the oil storage tanks were being filled appeared identical to those he had seen in Lourenço Marques: brown and grey wagons, with no markings on the sides.

Having confirmed this, Bernard then invented a pretext to meet the local head of Caltex for a drink. After a while Bernard gently brought the conversation around to oil. The Caltex man looked over the top of his pink gin, grinned, and said: 'Young man, if you even asked me what *colour* petrol is, I wouldn't be able to tell you. We have a ferocious Official Secrets Act.' Bernard, hardly surprised, concluded that Rhodesia was the one place where it was impossible – or too risky – to investigate sanctions-busting.

He returned to London with two facts confirmed: the main oil route was the railway from Lourenço Marques; and supplies were handled in some way by the South African firm of Freight Services Limited. Further research in the City Business Library showed that eighty per cent of the shares in Freight Services were held by two well-known companies: Anglo American and Charter Consolidated. The Anglo American group, controlled by Harry Oppenheimer, had tried to cultivate a 'liberal' image in South Africa, which made it most interesting to discover that one of its companies was handling Rhodesian oil. Charter Consolidated, which itself is partially owned by Anglo American, is a British company, registered in London. Two other UK companies also held small shareholdings in Freight Services: British and Commonwealth Shipping, and Ocean Transport and Trading. Already a British connection was appearing in Rhodesia's oil trade.

Bernard was uncomfortably aware that in answering what he had initially seen as the key question – the identity of Company X – he had merely opened up further issues that would be harder to resolve. Were the international oil companies unaware that the consignments they were selling to Freight Services were destined for Rhodesia? Or was it possible that Freight Services was acting on behalf of one or more of the oil companies? At this stage it seemed doubtful that the oil companies themselves would risk any involvement in sanctions-busting. This could, after all, have landed

23

their executives with a two-year prison sentence and an unlimited fine.

In Southern Africa there are five 'international' oil companies, the largest in the region being Shell. Shell's ownership structure is rather complex: the Shell Transport and Trading Company, registered in London, has a forty per cent stake in the Shell group, and the Royal Dutch Petroleum Company, based in The Hague, has the remaining sixty per cent. In Southern Africa Shell worked very closely with BP, in which the UK Government had a controlling shareholding. For most of the period since UDI, Shell and BP operated through jointly-owned subsidiaries in Southern Africa (held by Consolidated Petroleum), which were actually managed by Shell. Together the two British-owned companies controlled half of the market in Southern Africa – a heritage of Britain's colonial role.

Two American firms, Mobil and Caltex, also hold an important share of the market. Caltex itself is jointly owned by two of the largest US oil companies – Standard Oil of California and Texaco. And the final oil company in Southern Africa is the Compagnie Française des Pétroles (Total), in which the French Government has a forty per cent voting share.

It was back in London that the big break came. Bernard finally managed to meet someone who had worked for Mobil in Rhodesia. The Mobil man was extremely nervous about his identity coming out, and was always referred to as Oliver – a code-name deriving from 'olive oil'. Bernard met Oliver in secret on several occasions, always at different locations, and gradually the two men built up trust in each other. Oliver felt that Mobil was acting wrongly in supplying Rhodesia, but initially he was reluctant to reveal exactly how the company was busting sanctions.

Their fourth meeting took place at a tea house in the middle of the tiny Victoria Embankment Gardens next to Charing Cross station. In earlier meetings Bernard had revealed little of his own knowledge. But by this time he had realised that Oliver might be more inclined to provide information if he felt he was talking to someone who already understood some of the complexities of the affair. Bernard therefore asked: 'Tell me, does the name Freight Services mean anything to you?' Oliver looked up suddenly. 'Well,'

24

he replied slowly, 'it's known that they handle some trade for Rhodesia.' Bernard admitted that this was common gossip out in Southern Africa: 'But I'm talking about them acting as a front for the oil companies in Lourenço Marques.' Oliver suddenly began to open up. He explained that Mobil itself used Freight Services to supply Rhodesia, and added dramatically that he actually had secret documents to prove the link.

At this point Bernard and Oliver reached an impasse. Oliver had highly revealing and incriminating documentation which he wanted exposed to maximum effect – but he imposed two conditions. First, he naturally insisted that his identity should be protected. Secondly, his highly sensitive documents had to be handled by an organisation capable of analysing and publicising them to the full – but he was unwilling to trust the more obvious newspapers and television programmes for this task.

Bernard, enormously frustrated at not even being given a glimpse of the documents, eventually persuaded Oliver to meet a white South African exile who had in the past carried out some very effective work researching and exposing sanctions-busting. In addition, the exile had told Bernard that he was a member of the African National Congress (ANC) – South Africa's oldest liberation movement – and was authorised by them to carry out certain research projects. Bernard invited the two men to meet at a Wimpy Bar near Oxford Circus, but after a brief introduction was frustrated when it was hinted that it was time for him to leave. The two men talked alone. The meeting clearly went well, though, and afterwards Bernard was told that Oliver had agreed to hand over 95 pages of documentation which had been secretly copied in Mobil's Salisbury office.

It was to be a full year before Bernard finally saw the documents. During this time he learned that the South African exile had not told him the whole story of his relationship with the ANC. Instead of acting directly for the ANC, he in fact represented a secret group of white South African exiles named Okhela, which itself had certain rather unclear links with the ANC. Okhela, whose name came from the Zulu word for 'spark', was a clandestine organisation of white South Africans who provided what they called 'invisible support' to the ANC. But puzzlingly, when Okhela's existence was later publicly revealed, the ANC denounced them. The truth seems to have been that Okhela did at one time have links with some members of ANC, but that these were later closed. Certainly the ties between the two organisations had never been formal ones.

Shortly after the Mobil papers had been handed over to the ANC, things began to go badly wrong for Okhela. In August 1975 one of its leaders, the Afrikaner poet Breyten Breytenbach, was captured by the security police on a secret mission inside South Africa. The organisation was also plagued by all the problems and frustrations endemic to exile groups. Differences developed among Okhela activists, and the tenuous links with the ANC were severed.

The explosive Mobil papers stayed sitting in the ANC's files. Okhela, angry that the documents had not been released, decided to go ahead and publish them on their own. But there were two problems. First of all, the documents themselves consisted of a mass of technical papers and invoices which needed to be analysed and published with a commentary explaining their significance. Secondly, it seemed important that the report should be released by a respected organisation which could add its backing to the allegations. The Mobil documents were copies, not originals, and without access to the company's files in Salisbury it might well be difficult to prove that they were genuine.

In April 1976 the Reverend Larold Schultz, Director of the Center for Social Action of the United Church of Christ (UCC) in New York, received an overseas telephone call from a person claiming to represent Okhela. One month later a package containing the Mobil documents arrived by post. Okhela believed that the UCC, one of America's major Protestant churches, was a good channel through which to release the Mobil papers since its Center for Social Action had already been involved in an exposé of sanctions-busting. Two years earlier the Center had revealed that American airline companies were arranging ticket sales in conjunction with Air Rhodesia.

The Director of the UCC Center for Social Action was told that Bernard had already begun his own investigation of how Rhodesia imported its oil and was well qualified to write an analysis of the Mobil documents for publication. Bernard was therefore invited to New York, where he was able to study the secret papers for the first time. He was astonished. Not only did they confirm in considerable detail that Mobil subsidiaries had been deeply involved in a scheme to supply oil to Rhodesia via Freight Services, but they also contained some extremely candid remarks showing quite clearly that the arrangements had been set up with the deliberate intention of concealing Mobil's involvement in sanctions-busting.

The secret Mobil documents revealed that the company's Southern African subsidiaries had set up what they themselves called a 'paper-chase' to supply Rhodesia through a chain of intermediaries. One internal Mobil Rhodesia memorandum dealing with 'product procurement' explained its purpose: 'a carefully planned "paper-chase" is used to disguise the final destination of these products . . . This "paper-chase", which costs very little to administer, is done primarily to hide the fact that MOSA [Mobil South Africa] is in fact supplying MOSR [Mobil Rhodesia] with product in contravention of US Sanctions Regulations'.[1]

The Mobil documents also revealed that attempts had been made to disguise the Rhodesian trade by a series of rather clumsy code-words. An internal Mobil Rhodesia memorandum explained that aviation gasoline 115/145 was referred to as 'pineapples', and aviation turbine fuel as 'pawpaws'.[2] Freight Services documents later obtained from Mozambique showed that the code-letter 'A' represented Shell, 'AME' was Salisbury, and liquid petroleum gas was 'turnips'.[3] Clearly the person who had invented the code had a horticultural bent, and the staff who handled the Rhodesian trade derived some amusement from this amateur attempt at security. On one occasion a shipment of contaminated 'turnips' was described as being 'frothy-brown like the top of a glass of badly-poured Guinness.'[4]

Mobil's paper-chase for supplying Rhodesia involved two key intermediaries. First of all there was Freight Services, the South African-based forwarding company, which purchased oil supplies from Mobil South Africa. For obvious reasons the oil company was reluctant to sell directly to Rhodesia, and therefore had to supply their products through a company which would appear in their records as a normal South African customer.

The Mobil documents showed that Freight Services had

tried to disguise its involvement by establishing a number of 'paper' companies – which were often little more than letter-boxes for passing invoices. One of the most frequently-used fronts for Freight Services was a firm which went under the innocuous-sounding name of Minerals Exploration Limited. Freight Services, however, was at the centre of the spider's web.

The second main intermediary was a secret Rhodesian agency known as Genta – the name being an anagram of 'agent'. Genta had been set up in February 1966, shortly after sanctions had been introduced, to coordinate the importation of oil. Its role was to purchase fuel supplies through Freight Services, and then to distribute oil products to the Rhodesian subsidiaries of the international oil companies for local marketing. Although nominally a private company, Genta was actually owned by the Rhodesian Government. It was headed by George Atmore, formally a civil servant, who quickly established a position as the 'king' of the sanctions-busters.

The paper-chase and the use of intermediaries helped to disguise the incriminating links between the oil companies and Rhodesia. The basic system revealed by the secret Mobil documents was simple. Mobil South Africa sold oil to Freight Services, a 'legitimate' South African customer; Freight Services, since it was a South African company (and therefore able to trade with Rhodesia without fear of legal complications), would then sell the same consignment of oil to Genta; finally Genta would sell the shipment to Mobil Rhodesia for eventual distribution to the public. Payments would go back along the same channels.

The basic paper-chase was changed from time to time, and different procedures were used for the various oil products in order to make it more difficult to investigate the flow of oil. But one of the typical procedures revealed in the Mobil documents was the route used to sell lubricants to Rhodesia. Under this extremely complicated arrangement, Mobil South Africa, on receiving an order from Mobil Rhodesia, would sell a consignment of lubricants to a South African company known as Semco; the same shipment would then be passed to one of three companies – Rand Oils, Village Main Distributors, or Western Transvaal Development and Exploration – which in turn sold the consignment to another of these same firms; after passing through this complex channel, the lubricants went on to Minerals Exploration, which we have seen, was a front for Freight Services. On this occasion Genta did not handle the lubricants, since the government agency was not always used for the specialised oil products, and the Mobil consignment instead passed through a firm known as Recom of Rhodesia before finally reaching Mobil Rhodesia for local marketing. (The complete paper-chase is illustrated in Fig. 1.)

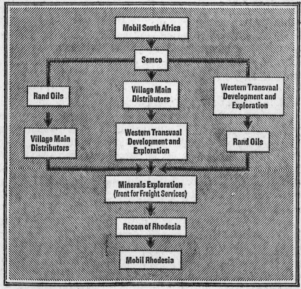

Mobil paper-chase for lubricants

Fig. 1

The complicated paper-chase used by Mobil South Africa to sell lubricants to Mobil Rhodesia. Semco, on receiving lubricants from Mobil, would supply them through three different routes to disguise the sale. After passing through five 'paper' companies, the lubricants would finally reach Mobil Rhodesia.

It should be stressed that this complicated procedure was only used for the paper-work, the shipments of oil themselves being normally railed directly from Lourenço Marques into Rhodesia. Mobil had, however, made one fatal error. In an attempt to keep the efficient office procedures, the company's Rhodesian subsidiary had put the whole thing on paper in an internal memorandum.

Bernard, who was absolutely astonished at the detail and incriminating nature of the material contained in the Mobil papers, worked extremely hard, and quickly, to analyse and write up the material for the United Church of Christ. The final decision to be made was the title of the pamphlet. Eventually it was called *The Oil Conspiracy* – a title which became more and more appropriate when the full complicity of Western companies and governments in the sanctions scandal later emerged.

* * * *

The Oil Conspiracy was launched at a press conference in Washington on 21 June 1976. The timing had made the issue particularly topical, and extensive press publicity was expected. Just five days earlier, demonstrations by angry blacks had broken out in the South African township of Soweto, and two days afterwards US Secretary of State, Dr Henry Kissinger was due to hold crucial talks with South African Prime Minister John Vorster in Germany. International attention was riveted on Southern Africa.

The press conference was jointly sponsored by the United Church of Christ and a new organisation known as the People's Bicentennial Commission. The Commission, which was led by former anti Vietnam war activists, had set out to publicise corporate abuses as an antidote to the official American Bicentennial celebrations. A central part of the People's Bicentennial Commission's campaign was the offer of a $25,000 reward to anyone who could produce evidence of a major criminal act by an American company leading to prosecutions. In announcing the press conference to launch *The Oil Conspiracy*, the Commission tempted journalists to attend by announcing that there would be an exposé of one of America's ten largest corporations. No mention was made, however, of either Mobil or Rhodesia. The appetite of the press was whetted: a hundred journalists and three TV crews crowded into the Washington press conference.

The People's Bicentennial Commission's reward was in two parts: $10,000 for evidence received of a corporate criminal act, and a further $15,000 if conviction followed. The cheque for the first $10,000 instalment was sent to the United Church of Christ immediately after the press conference. Two weeks later the money was returned by the UCC on the grounds that publication of *The Oil Conspiracy* was part of its on-going work and should require no special reward. But behind the scenes some members of the board of UCC's Center for Social Action were critical of the decision to publish the report, and objected to the fact that it was based on unauthenticated documents, particularly since the documents had been obtained from an underground group (Okhela) and publicised with the help of the radical People's Bicentennial Commission.

After the press conference, *The Oil Conspiracy* sank like a lead balloon. Newspaper coverage was almost non-existent. *The New York Times* ran nothing, *The Washington Post* carried only a brief story buried well into the business section, and there was no coverage in *Time* or *Newsweek*. The allegations were not even mentioned on the television news. What had gone wrong? *Time*'s Don Sider claims that he did a story, but that the editorial department decided not to run it. 'They were highly exciting charges,' he later admitted, 'but I also filed a strong caveat with my editors – the documents we were given at the press conference were complex, and we had no way of knowing whether they were real or not. I invested several hours in the story, talked with the Justice Department and others, but my heart just wasn't set on it, and we just don't have a month for tracking undocumented charges.'[5]

The simple fact of the matter was that American journalists were not really interested in Rhodesia or sanctions, and that British reporters did not take up the story because the documentary evidence related to a US company.

Mobil's initial reaction to *The Oil Conspiracy* was one 'of shock, of disbelief,'[6] and they retaliated with the old trick of pleading innocent to crimes of which they had not been accused. They denied that Mobil had broken US sanctions legislation, and also maintained that their South African subsidiary had not sold oil to Rhodesia. But both claims avoided the central charge: that Mobil South Africa

31

was supplying Mobil Rhodesia *via Freight Services*.

The day after the press conference, Mobil's headquarters reported that William Beck, Chairman of Mobil South Africa, had assured his New York head office that the requirements of US sanctions law had been 'scrupulously observed'.[7] American sanctions legislation only applied to companies registered in the United States, not to subsidiaries incorporated abroad, and from a legal point of view, Mobil South Africa was therefore quite free to trade with Rhodesia. But US sanctions legislation also made it an offence for any American citizen to be involved in sanctions-busting, or for oil products manufactured in the United States to be exported to Rhodesia. It was therefore still possible that Mobil was breaking US sanctions law.

Mobil immediately initiated its own internal investigation into *The Oil Conspiracy* charges. A request for further information was sent to William Beck, who replied that Mobil South Africa was prevented from assisting by the local Official Secrets Act. Mobil's US management therefore decided that an on-the-spot investigation was necessary. On 5 August 1976 George Birrell, Vice-President of Mobil, arrived in Johannesburg. He immediately took legal advice, and was warned by his South African lawyers that if the Mobil visitors attempted to investigate the flow of oil to Rhodesia 'they themselves would be subject to prosecution as foreign agents under the Official Secrets Act'. The lawyers added that Rhodesian laws were even stricter, and that the Mobil men 'would run a substantial risk of arrest if they were simply to enter that country for this purpose'.[8]

In other words, according to this advice, a top manager of one of America's largest corporations was in danger of being arrested as a foreign spy if he merely attempted to *inquire* into the affairs of a wholly-owned subsidiary in South Africa. This was an astonishing situation. It also made a mockery of Mobil's argument that by continuing to operate in South Africa the firm had a 'liberalising' effect on the apartheid regime.

It later emerged, however, that Mobil might well have been using the existence of the Official Secrets Act to conceal its sanctions-busting activities. Mobil never explained why its investigation team did not go on from South Africa to neighbouring Mozambique. After all, until earlier that year

Rhodesia's oil supplies had passed from South Africa through the port of Lourenço Marques, and there was no similar secrets act in Mozambique to prevent Mobil from pursuing inquiries there. Fernando Gomes, Mobil's manager in Lourenço Marques, would undoubtedly have been able to provide full details of the clandestine shipments which had been made to Rhodesia, had anyone bothered to ask him. A visit to Portugal would also have enabled the investigators to have interviewed retired Portuguese staff who had served with Mobil in Mozambique. No wonder, then, that one of the Rhodesian nationalist organisations went as far as to suggest that 'a group of boy scouts would have done better' than the Mobil executives did in trying to investigate the charges.'

Mobil's most detailed reaction to *The Oil Conspiracy* allegations was presented to Senator Dick Clark before the Senate Subcommittee on African Affairs on 17 September 1976. This included a 38-page testimony by Vice-President George Birrell, a 57-page report by Mobil's lawyers, and over 400 pages of supporting documentation. One informed source has claimed that it cost Mobil one million dollars to prepare this defence – a considerable sum to invest, bearing in mind that *The Oil Conspiracy* charges had been virtually ignored by the press. Was Mobil protesting too much? Or did they recognise the seriousness of the allegations and consider it worth burying them once and for all?

Mobil's defence before the Senate was important, because the same arguments were later to be deployed by the British oil companies when Shell and BP faced similar charges. Mobil's carefully designed defence rested on two main points. First, it was claimed that the Official Secrets Act in South Africa made it impossible to examine allegations that its South African subsidiary was a clandestine supplier of Rhodesia; secondly, that even if these charges were true, nothing could be done to stop this trade, since South African legislation on 'conditional selling' made it illegal for the oil companies to refuse to sell to any customer willing to pay the market price.

Birrell told the Senate Subcommittee on African Affairs that Mobil had been prevented by the Official Secrets Act from authenticating the documents reproduced in *The Oil Conspiracy,* and went on to suggest that the documents

33

might well have been forged. The allegations, Birrell pointed out, could possibly be part of a 'skilfully contrived publicity scheme' or 'an artful blend of fact and fiction' designed to 'hoodwink respected institutions' like the United Church of Christ and the US Senate.[10]

Mobil did not, however, produce any substantial evidence to show that the papers had actually been forged. Surely, if any of the seventeen long documents reproduced in *The Oil Conspiracy* had been forged, then Mobil would have been able to point to at least some details which suggested that they were not genuine? And, besides, why should someone who wanted to implicate the company in sanctions-busting produce such complicated forgeries when simpler and more clearly damning documentary evidence would have done the job just as well?

Birrell maintained that strenuous efforts had been made to examine the authenticity of the charges, but that the Official Secrets Act had proved an insurmountable obstacle. He did admit, however, that this had been the first occasion on which Mobil South Africa had invoked the Act in refusing to pass on information. This suggests that Mobil was either using the Official Secrets Act as an excuse – or else that the company had never before made a serious attempt to establish whether its South African subsidiary was observing sanctions.

At one point during the Senate hearings Birrell described how he had tried to contact a Mobil employee in South Africa, Nicholas Gubb. First of all he had looked up Gubb in the telephone directory. But this had led nowhere: 'I was unable to locate him. My information is that he has recently moved from Cape Town to Durban. He has no telephone'. Senator Clark suggested the simple expedient of phoning his office: 'I thought he worked for Mobil?' Birrell replied: 'He does, but . . . we know that our South African affiliate will ask us, particularly me, why I want to talk to him. The minute I tell them, they will not permit him to speak to me.'[11] In other words, Mobil's Vice-President was apparently not even permitted to speak to an employee of the company's wholly-owned subsidiary in South Africa. Just a year later, however, Shell and BP were able to ignore similar warnings and obtain a considerable amount of information about their South African operations.

In effect Mobil was suggesting to the Senate that it had virtually lost control over its operations in Southern Africa. Technically Mobil's New York head office had no authority over its subsidiary in Rhodesia, because after UDI the oil companies had been 'directed' by the Rhodesian regime not to accept orders from their parent head offices abroad. But Birrell certainly misled the Senate by giving the impression that the head office in the United States had so little authority over its South African firm. The fact was that the Board of Directors and the Chairman of Mobil South Africa were appointed by the American parent company.

Mobil's second line of defence was to claim that even if *The Oil Conspiracy* charges were correct – and this had not been proved – then there was nothing that could be done, since South African legislation on 'conditional selling' made it an offence for Mobil South Africa to attach conditions to oil sales, or to refuse to sell to any customer who was willing to pay the market price. The company was therefore quite unable to cut off sales to any South African firms who might subsequently be exporting to Rhodesia.

At the time this argument was not seriously questioned by the Senate. But again there was an apparent inconsistency in Mobil's testimony. The company's South African lawyers had argued that the 'conditional selling' legislation applied only to legitimate South African customers. 'It is our under-standing,' they advised, 'that Mobil South Africa remains legally free to decline to sell to any purchaser who is obviously a Rhodesian entity, or to anyone who is plainly acting as a mere agent for a Rhodesian purchaser".[12] All the evidence pointed to the fact that Freight Services was an agent for Rhodesia, and that Mobil could therefore have cut off sales to the forwarding company. Indeed this was precisely what BP actually did a few months later. In July 1977 BP South Africa halted all supplies to Freight Services, and no legal action was taken by the South African authorities.

But before the Senate at least, Mobil had appeared to rebut *The Oil Conspiracy* charges successfully. The Reverend Larold Schultz, Director of the UCC Center for Social Action, later recalled the scene: 'Ten top executives of Mobil arrived in a phalanx, all in blue striped suits, with a 400-page document, while there were only three representatives from

35

the UCC.'[13] In the end the Reverend Schultz proved no match for the well rehearsed legal arguments from Mobil.

At this stage, one oil expert concluded, the allegations made in *The Oil Conspiracy* were now 'a matter to file and forget.'[14] But the Mobil case had rested on a highly questionable interpretation of South African law. Now that further facts have been revealed, suggesting that Mobil's defence is untenable, the company may one day regret its submission to the prestigious US Senate.

Meanwhile, the American Government continued to investigate *The Oil Conspiracy* charges through the usual channels. Sanctions enforcement in the United States is the responsibility of a department known by the cumbersome title of the Office of Foreign Assets Control of the Treasury. On 30 June 1976, just nine days after publication of *The Oil Conspiracy*, an order was served on Mobil's New York headquarters to produce documentary material relating to Southern Africa. Subsequently both the State Department and the Secret Service were called in to assist the Treasury investigation.

The Treasury inquiry took nearly a year, and its report was finally released on 17 May 1977. It concluded that, with two exceptions, *The Oil Conspiracy* documents 'could not be authenticated', because the originals in the files of Mobil Rhodesia could not be consulted. 'Each document is no more than a written statement or communication which may be what it purports to be or may, in fact, be a forgery.'[15] The Treasury accepted Mobil's legal defence, and the findings of their inquiry were completely inconclusive. There was no positive evidence that American law had been broken.

It is now clear that the Treasury investigators lacked a certain enthusiasm for discovering the truth, for an investigation initiated by Senator Edward Kennedy, on the Implementation of Economic Sanctions Against Rhodesia confirmed that US sanctions enforcement had been lax. This study, which was prepared by the Comtroller General, was released just six days after the Treasury report on Mobil. It concluded that there was a 'lack of emphasis on fully enforcing the UN sanctions' by American government agencies, and that this was partly due to 'the low priority assigned to this activity'.[16] Senator Edward Kennedy commented that 'the United States has not been living up to its public commitments on enforcing sanctions against Rhodesia'.[17]

* * * *

Bernard Rivers was very disappointed with the response to *The Oil Conspiracy*. Undoubtedly the Mobil affair was the most important case of sanctions-busting to have been exposed since UDI. Yet through legal sophistry Mobil appeared to have succeeded in rebutting the charges against them. It was not until later that the crucial impact of *The Oil Conspiracy* in uncovering the sanctions scandal became clear.

NOTES

1 Mobil Rhodesia memorandum, *The Oil Conspiracy*, p. 39.
2 Mobil Rhodesia memorandum from D. Wilson to Stephen Phear, 2 March 1971 (unpublished).
3 Letter from Dan Airey (Genta, Salisbury) to Jock Davidson (Freight Services, Lourenço Marques), 6 July 1972 (unpublished).
4 Telex from Dan Airey (Genta, Salisbury) to Jock Davidson, 11 August 1972 (unpublished).
5 *Mother Jones*, September–October 1976, p. 22.
6 Testimony by George Birrell to US Senate, 17 September 1976, p. 30.
7 Mobil statement, 22 June 1976.
8 Mobil statement, 17 September 1976.
9 Memorandum on Sanctions Violations, submitted by United African National Council to US Senate, 10 March 1977, p. 10.
10 Mobil statement, 17 September 1976.
11 George Birrell's oral testimony to US Senate, 17 September 1976.
12 Report by Edward Fowler (General Counsel of Mobil), 10 September 1976, p. 56.
13 *Sunday Telegraph*, 10 September 1978.
14 Jorge Jardim, *Sanctions Double-Cross*, p. 128.
15 Treasury Investigation of Charges made against the Mobil Oil Corporation, p. 33.
16 Implementation of Economic Sanctions against Rhodesia, 20 April 1977, p. iii.
17 *Evening Standard*, 23 May 1977.

2 Shell Shock

Although *The Oil Conspiracy* report dealt primarily with Mobil, it did contain a very short section on the other international oil companies. Shell, in particular, was singled out for a special mention. The few lines concerning Shell later proved of vital importance in exposing the key role played by British oil companies in supplying Rhodesia.

Work on preparing *The Oil Conspiracy* had been conducted in great secrecy. Although Bernard was a very close friend of mine, he had never mentioned his meetings with Oliver. He did tell me, however, that he had been investigating how Rhodesia got its oil. I myself was interested in Southern Africa, and quickly became bitten by the sanctions story. After completing a doctorate on Tanzanian foreign policy at the London School of Economics, I had gone into freelance journalism, specialising in African affairs. I could see that an exposé of how Rhodesia was defying the UN oil embargo would be a major scoop, and like Bernard, I believed that publication of the facts would lead to further pressure to tighten sanctions against the rebel regime.

Soon after Bernard told me about his interest in Rhodesia's oil, I had the good luck to meet a Shell executive who knew about the company's operations in Southern Africa. Bernard encouraged me to develop this lead, and began to tell me rather more about the trail that he had been secretly pursuing over the previous year. One lunchtime I arranged to meet my Shell contact over a pint of beer at a King's Cross pub. It was here that I first learned of the crucial involvement of the British oil companies. Right from the start my informant was nervous; but as I brought the conversation around to Rhodesia he became increasingly ill at ease. I told him that I had recently received information suggesting that a major international oil company was

38

supplying Rhodesia through intermediaries. 'Is it Shell?' he asked immediately. No, I told him, it was an American firm. Then after further questioning, he admitted that Shell South Africa was indeed selling oil to Rhodesia via third parties.

I went on to ask whether Freight Services was the main intermediary. He quickly nodded. This admission provided a crucial link in the story. Bernard had heard about Freight Services out in Lourenço Marques, and its use by Mobil had later been confirmed by Oliver. But at this stage neither of us had yet seen the incriminating Mobil documents. We were therefore fascinated to discover that Freight Services was also handling supplies for the subsidiary of a *British* oil company.

My Shell informant appeared to be talking from detailed knowledge, not mere gossip. He went on to add that the sales records of Shell South Africa included a category enigmatically entitled 'FS' – which stood for Freight Services. Every three months, he explained, Freight Services would contact Shell to specify their requirements of different oil products over the next three and twelve months. Most important, however, was the fact that Shell South Africa apparently knew exactly what happened to the oil it sold to Freight Services. It was not as if Freight Services was just another South African customer; 'FS' meant Rhodesia – and Shell staff knew it.

The details revealed by my Shell informant were used by Bernard when he later wrote *The Oil Conspiracy*. There were also incidental references to Shell scattered throughout the secret Mobil documents. We were fascinated to find that the occasions when Shell was mentioned usually occurred when the British-owned company was apparently trying to grab part of Mobil's share of the Rhodesian market. For example: a confidential Mobil Rhodesia memorandum on 'Product Procurement' discussed procedures for importing various aviation fuels. It pointed out that aviation gasoline 100/130 was 'imported from Shell' and added that aviation turbine fuel was supplied by Mobil 'despite frequent attempts by Shell to stop this'.[1] Another letter from Mobil Rhodesia to Mobil South Africa also referred to further attempts by Shell to increase their sales: ' . . . we [*i.e.* Mobil] have held further discussions with Genta . . . and they are aware that we are going back to Shell with an offer to accommodate them at LM [Lourenço Marques] for Regular. If Shell refuse this offer there will be no justification for Genta favouring them on Premium at our expense.'[2]

A further Mobil letter, written at the height of the 1973 oil crisis, admitted that 'Shell draw their hexane from Abadan [in Iran] and are probably more vulnerable than we are to supplies being cut off.'[3] The Mobil documents also included a number of references to the use of Shell and BP subsidiaries as intermediaries in supplying Rhodesia. African Bitumen Emulsions, a company in which Shell has a fifty-four per cent shareholding, apparently handled Rhodesian imports of industrial bitumen. Semco, in which Shell and BP together had an indirect thirty-five per cent stake, was used as an intermediary for lubricants and greases.

Oliver, the source of the Mobil documents, had also told Bernard about a new Shell lubricant blending plant which had been opened in 1974 at Willowvale, on the outskirts of Salisbury. *The Oil Conspiracy* reported that the 'base-stock' (semi-processed crude oil) which was used to manufacture the lubricants 'comes from South Africa, all or nearly all from the Shell refinery'. Apparently the plant not only supplied Shell service stations in Rhodesia, but the entire market. The lubricants were blended according to Shell specifications and then put into tins marked with the trademarks of Shell, BP, Mobil, Caltex, and Total. Shell therefore had a virtual monopoly on the production of this vital oil product.

The Oil Conspiracy contained considerable circumstantial evidence that Shell South Africa was deeply involved in sanctions-busting. Admittedly, the source of our information about the use of Freight Services was not named in the report and the secret Mobil documents only mentioned Shell in passing. Yet the facts, when put together, certainly suggested that British-owned companies could well be Rhodesia's major source of oil. Bernard and I therefore hoped that the publication of the report would force the issue into the open in Britain.

When *The Oil Conspiracy* report was released in June 1976, we distributed copies to the British press with a special note pointing out the references to Shell and BP. We were disappointed, however, to find that the report generated just as little interest in Britain as it had in the United States.

40

Like Mobil, the British companies claimed there was no basis to the charges.

Two weeks after the publication of *The Oil Conspiracy*, BP denied that it had been sending oil into Rhodesia from Lourenço Marques. *Platt's Oilgram*, the oil trade newspaper, reported that 'BP said it doesn't trade with Rhodesia and has fully complied with the UN sanctions.'[4] The BP statement, it later emerged, was quite untrue. *Platt's Oilgram* added that BP had not had any contact with its Rhodesian company since UDI. Yet it later emerged that only two years earlier three British BP representatives, (Denys Milne, Alan Robertson, and Neil Webber) had actually visited Salisbury for discussions with the General Manager of the Rhodesian marketing firm (B12.8).

Shell gave their reaction to allegations of sanctions-busting to The *Sunday Times*: 'Shell denied, for the umpteenth time, that its South African subsidiary had knowingly supplied these products to Rhodesia. Hans Pohl, Managing Director and Acting Chairman of Shell Oil South Africa, said: "As far as I know no Shell company in South Africa has an interest in supplying oil to Rhodesia".'[5] Pohl's reply appears to have been extremely misleading, for the following year he admitted that he had learned back in 1971 that 'there were certain arrangements being made via a company in South Africa, Freight Services, and by this means Rhodesia was supposed to get . . . from the sources of Shell and BP, its petroleum products' (B8.29).

The British oil companies continued to respond to *The Oil Conspiracy* with a deliberate attempt to cover up the facts. In answer to a letter I wrote asking for more detailed comments on allegations that their South African subsidiaries had supplied Rhodesia through Freight Services, Shell replied: 'No company in which Shell has an interest supplies oil to Rhodesia'.[6] BP also stated that it 'does not trade with Rhodesia,'[7] I immediately wrote back asking for answers to my specific questions about sales through Freight Services. Neither company replied.

* * * *

At this time Bernard and I had not yet appreciated a crucial legal point concerning the handling of oil at the Mozambican port of Lourenço Marques. UK sanctions legislation made it an offence for any *British* firm or person to supply goods 'knowing or having reasonable cause to believe that they will be supplied or delivered to or to the order of a person in Southern Rhodesia', or performing 'any act calculated to promote the supply of or delivery' of such goods to Rhodesia.[8]

Over many years both Shell and BP had gradually incorporated their subsidiaries throughout Africa as local companies, for reasons quite unconnected with the Rhodesian situation. This had already occurred many years earlier in South Africa, where the Shell and BP marketing companies were registered in Cape Town. The South African subsidiaries of Shell and BP did not therefore break British law by trading with Rhodesia.

But there was one major difficulty: the jointly-owned Shell/BP company in Mozambique, known as Shell Mozambique, was registered in London, and its board was largely composed of British citizens. This meant that the Mozambican subsidiary of Shell and BP *was* subject to the UK sanctions law. But why was it necessary for Shell Mozambique ever to become involved in handling Rhodesian trade? The answer is that until 1974 there was no direct rail link between South Africa and Rhodesia. This meant that the most economical route for supplying the rebel regime was by rail from the Mozambican capital of Lourenço Marques. Shell and BP sales to Rhodesia were actually organised by the South African subsidiaries of the oil companies. But the oil was physically sent through Lourenço Marques, and passed through the installations of Shell Mozambique. While this happened Shell Mozambique, as a British company, was clearly committing an offence against the UK sanctions law.

At the time of the publication of *The Oil Conspiracy* neither Bernard nor I was aware that Shell Mozambique was a British-registered firm. But this simple fact was to be of central importance.

*   *   *   *

Meanwhile *The Oil Conspiracy* report was formally presented to the United Nations. On 2 July 1976 the UN sanctions Committee met to discuss the 'Supply of Oil and Oil

Products to Southern Rhodesia'. The case was numbered INGO-17 (INGO signifying that it had been raised by an individual or non-governmental organisation), and it was to become the most important of the four hundred-odd alleged breaches of sanctions which have been reported to the United Nations. Despite the fact that oil was absolutely crucial to Rhodesia's survival, and the country was clearly obtaining regular supplies of it, this was actually the first occasion since its establishment that the UN Sanctions Committee had considered an alleged breach of the oil embargo. It is also interesting that even then the case was not reported by a government. All this reflected badly on the policing roles both of member governments and the United Nations itself in enforcing sanctions against the rebel regime.

The Reverend Donald Morton, then a leading member of the US Anti-Apartheid Movement, formally presented *The Oil Conspiracy* to the United Nations. After explaining the significance of the Mobil paper-chase, he suggested that 'the role of Shell in Rhodesia was at least as important as Mobil and should be investigated fully'[9] The UN Sanctions Committee then officially asked the British and Dutch Governments to investigate the allegations.

On 2 September 1976 the British Government sent an official reply to the UN in which the following astonishing claims were made:

> The competent United Kingdom authorities have studied the report most carefully, and have discussed its contents with the British oil companies mentioned. These authorities are satisfied that the report contains no evidence of sanctions-breaking by any British companies or individuals, and have accepted the assurances given by Shell and BP that neither they nor any company in which they have an interest have engaged either directly or with others in supplying crude oil or oil products to Rhodesia. This is the same position as that established in 1968, when Her Majesty's Government investigated similar charges at the highest level with the same companies.[10]

Bernard and I were confused. The British letter, after all, was not an off-the-cuff remark but a carefully formulated official statement. It very specifically denied that British oil companies or their subsidiaries had been 'engaged either directly or with others'. Surely this didn't mean that we were better informed than the British Government? Or would the British Government actually lie to the United Nations? The official version seemed to be at such variance with our own that for a moment we wondered whether we might have got our story wrong.

It later transpired that the British statement was quite untrue. We learned that Shell and BP had been careful *not* to give the precise 'assurances' indicated by the Government. Shell and BP subsidiaries *had* for a number of years been engaged – sometimes 'directly', on other occasions 'with others' – in supplying Rhodesia. The position, it later emerged, was emphatically *not* the same as that 'established in 1968'.

*   *   *   *

It was only recently that Bernard and I were able to piece together the full story of how Britain had so completely misled the United Nations. In 1974, John Francis, a Shell official visiting South Africa, had discovered that the jointly-owned Shell-BP company in Mozambique was illegally supplying oil to Rhodesia. The news was then passed back to the executive responsible for Southern Africa at Shell Centre in London. But for various reasons this embarrassing discovery was apparently not relayed up the hierarchy of the company, and the Chairman was not informed. The result was that Shell misled the British Government – and the first link in the chain of misinformation was forged.

On 30 June 1976, Sir Frank McFadzean, the Chairman of Shell, wrote to the Foreign Office to comment on *The Oil Conspiracy* allegations. This was in fact Sir Frank's last day at Shell before he took up his new appointment as Chairman of British Airways. His letter began by referring to the subsequently notorious 'swap' arrangement with Total, which had been introduced back in January 1968. Under this swap, the requirements of Shell and BP in Rhodesia were supplied by the French company of Total, and in return Shell and BP provided matching amounts of oil for Total's customers in South Africa. This cosmetic device enabled Shell and BP to continue to supply Rhodesia, but removed Shell Mozambique from involvement in actually handling the oil. Surprisingly the British Government condoned this dubious scheme. The Total swap remained a well-kept secret for more than a decade, and when *The Oil Conspiracy* was published, knowledge of the arrangement was confined to a mere handful of oil company executives, senior civil servants, and politicians.

After a brief reference to the 1968 swap arrangement, Sir Frank McFadzean went on to tell the Foreign Office in his June 1976 letter that 'nothing has happened since which fundamentally alters the realities of the situation as then outlined . . . no company in which we have an interest is supplying to Rhodesia' (B13.18). Sir Frank may have been unaware that his statement was quite untrue – although it was surely his responsibility to ensure that he was properly briefed. But the letter was actually drafted by John Francis, the Shell official who *had* learned two years earlier that the Total swap had been ended, and that Shell Mozambique had resumed handling oil for Rhodesia.

Just three months before the publication of *The Oil Conspiracy* an important change had taken place in the route used to supply Rhodesia. In March 1976 the newly-independent Mozambique Government had closed its border with Rhodesia, making it impossible for the oil companies to send oil directly from Lourenço Marques. From then onwards supplies were sent straight from South Africa into Rhodesia. As Sir Frank McFadzean told the Foreign Office, the border closure had removed Shell Mozambique 'as a source of leakage' (B13.18). But the fact is that *The Oil Conspiracy* had not even mentioned the existence of Shell Mozambique – it had alleged that Shell South Africa was selling oil to Rhodesia. Once again the oil companies were pleading innocent to crimes of which they had not been accused.

The Shell letter was sent to Sir Michael Palliser, head of the Foreign Office, who was himself no newcomer to the oil sanctions story. Back in 1968, at the time when the Total swap had been introduced, he had served at Number Ten as one of Harold Wilson's private secretaries. He had therefore received reports of the meetings between the Government and the oil companies at which sanctions had been discussed in detail. In his new job at the Foreign Office he should

45

therefore have been particularly well placed to understand the new allegations against Shell and BP.

When Sir Michael Palliser received the Shell letter he assumed that the company was continuing to receive orders from Rhodesia, but was meeting them according to the swap arrangement with Total. Sir Michael therefore replied to the new Shell Chairman, Michael Pocock, on July 9 1976: 'I welcome Sir Frank's assurances . . . We shall continue vigorously to defend the record of our oil companies in relation to Rhodesia when it comes up for discussion in the UN Sanctions Committee' (B13.19).

This was just what the Foreign Office did – with the official statement presented to the United Nations. What is, perhaps, most remarkable about the official reply is that John Francis, Sir Frank McFadzean, and Sir Michael Palliser all had different beliefs about what was actually happening: but the British statement to the UN did not accurately reflect *any* of these beliefs, true or false.

BP was in an even more sensitive position than Shell since the Government had a shareholding in the company and appointed two Directors to sit on its Board. Charges of sanctions-busting by BP were therefore particularly damaging, since it was the British Government itself which had first gone to the United Nations to propose the introduction of the oil embargo after UDI. On 1 July 1976 the BP Board of Directors discussed *The Oil Conspiracy* allegations and Shell's letter to the Foreign Office. Christopher Laidlaw, the Managing Director responsible for Africa, thought that Sir Frank McFadzean's letter was 'an admirable one' (B13.22), apparently believing that the Total swap arrangement had remained in force. He later claimed that at the time he was unaware that the jointly-owned Shell-BP company, Shell Mozambique, was registered in London and therefore prohibited from trading with Rhodesia. Laidlaw, who after all was the BP Director responsible for Africa, seems to have displayed considerable ignorance about the Rhodesian affair. Michael Savage, the BP Regional Co-ordinator for Southern Africa, seems to have had almost as little understanding of the situation. He too was shown a copy of Sir Frank McFadzean's letter, and he then consulted BP's Rhodesian files. Even though he had learned that 'Freight Services' was used as a 'pseudonym' for Rhodesian trade, he

46

was unaware that the Total swap arrangement had been ended several years earlier.

Like their counterparts in Shell, BP officials in London therefore seem to have believed that they were not involved in breaking the law. On 8 July 1976 Michael Savage visited the Foreign Office in order to discuss the situation with two senior officials responsible for Rhodesia (Aspin and Barlow) and to proclaim BP's innocence (B13.23). The British Government eagerly seized upon these 'assurances' from Shell and BP, and told the United Nations that Britain's hands were clean.

*　　*　　*　　*

The British Government was not the only body approached by the UN Sanctions Committee and asked to comment on *The Oil Conspiracy* allegations. Because Shell was a joint Anglo-Dutch company, the Netherlands was also asked to submit a report on the charges. Royal Dutch Petroleum, the Netherlands-based part of the Shell Group, was approached. But it was not until nearly a year later, on 10 March 1977, that the results of the Dutch investigation were presented to the UN:

> The Netherlands Government approached the Board of the Royal [Dutch] Shell Group . . . The Royal Shell Group had no knowledge whatsoever of any supply of oil or oil products to Southern Rhodesia in which Shell South Africa was involved.[11]

At the time Bernard and I had no reason to question this statement. But it now appears that the Dutch response was quite misleading: in fact for most of the period since UDI the post of Shell's Regional Co-ordinator for Africa had been held by a Dutch citizen. Dirk de Bruyne had not only been the Shell Regional Co-ordinator for Africa in the crucial years between 1965 and 1968. From 1972 until 1976 he was also the Managing Director responsible for Africa, and his Regional Co-ordinator was another Dutch executive, J. de Liefde. De Bruyne had therefore spent much of his time dealing with the tricky problems that developed over Rhodesia, and was present in person at many of the key

47

meetings at which supplies to Rhodesia were discussed. At the time of the Dutch statement to the United Nations, in March 1977, de Bruyne was a Managing Director of Royal Dutch in The Hague. Less than four months later he was promoted to head the company. Bearing in mind de Bruyne's close involvement in supervising Shell's trade with Rhodesia, it is difficult to understand how the Netherlands was able to tell the UN that the Board of Royal Dutch Petroleum 'had no knowledge whatsoever about oil supplies to Rhodesia'.

By the time of the Dutch Government's reply to the UN, Shell's involvement in Southern Africa had become an important political issue in Holland. Kairos, a church group concerned with Southern Africa, had already been involved in three years of discussions with the management of Royal Dutch Petroleum. Cor Groenendijk, who represented Kairos at these meetings, recalled that 'Shell responded to our questions about sanctions with nothing more than bland denials.[12] Indeed, in the summer of 1974 Royal Dutch Petroleum had written to Kairos to tell them that 'no company in which Shell has an interest supplies oil to Rhodesia . . . we cannot legally trade with or make investments in companies in Rhodesia.'[13] In 1976 Kairos decided to publish their findings on Shell's involvement in Southern Africa. Sami Faltas, a researcher working for the Amsterdam-based Ecumenical Study and Action Centre was commissioned to write a report, *Shell in Zuid-Afrika*, which by chance was published in Holland within a few days of the release of *The Oil Conspiracy*. But the allegations of sanctions-busting contained in the Dutch report were imprecise, and were immediately dismissed by Royal Dutch Petroleum.

The publication of *Shell in Zuid-Afrika* in Holland and *The Oil Conspiracy* in the US sparked off a general interest in the role of the oil companies in Southern Africa which resulted in my being commissioned to write a report on Shell and BP for the Haslemere Group and the Anti-Apartheid Movement. The Haslemere Group, which was concerned with Third World issues, had become particularly interested in the role of British economic interests in Southern Africa, and later helped to sponsor further research on the oil companies by Bernard and myself.

The Haslemere Group/Anti-Apartheid report *Shell and*

BP in South Africa was eventually published on 1 March 1977. Though mainly concerned with the operations of the oil companies in South Africa, it also contained a short section on Rhodesia, and it was this that attracted the most attention. The information itself was hardly new, since it had already appeared in *The Oil Conspiracy*, but this was the first occasion that this material had been published in Britain. After discussing documented evidence against Mobil, the report revealed that 'strong evidence is now emerging to suggest that Shell may also have been involved in selling petroleum to Rhodesia.'[14] It pointed out that there was little doubt that Shell and BP oil had been reaching Rhodesia – but remained cautious as to whether the South African subsidiaries of the two companies were themselves involved in this trade. The conclusions of *Shell and BP in South Africa* now appear very mild in the light of information which has surfaced since.

The day after the report on Shell and BP was published, the allegations were raised by angry backbenchers in the House of Commons. But the Government did not appear to take the matter particularly seriously. Ted Rowlands, Minister of State at the Foreign Office, warned Parliament that 'there have been occasions in the past when evidence that has been brought forward on this score has not been proved sound enough to take action'. He added that he was 'generally satisfied with the observance of sanctions by United Kingdom firms . . . our record in this respect is second to none.'[15] In fact the Foreign Office appeared to take little interest in evidence of sanctions-busting. My own efforts to meet a senior official to discuss evidence against the British oil companies were rebuffed for many months.

The oil companies themselves continued to deny any involvement in sanctions-busting.

Shell Centre issued a statement claiming that 'Shell companies have not been, nor are they, in breach of the UK Second Order in Council of 1968 dealing with sanctions on Rhodesia'.[16] The *Financial Times* was told by Shell that the company was not involved in sanctions-busting. But on this occasion the spokesman also claimed that it was not possible to say whether Shell South Africa supplied Freight Services, and added that the company could be liable to prosecution under South African law if it refused to sell to its customers.[17] Shell, in other words, had adopted the Mobil 'defence': the head office of the company knew nothing about sales made by its South African subsidiary – and in any case, under South African law it could not refuse to supply 'suspicious' purchasers. Shell even felt confident enough to brand its accusers as troublemakers. Two months later the Chairman, Michael Pocock, told shareholders at the firm's AGM that allegations about sanctions-busting came 'very close to sanctimonious humbug'.[18]

BP also dismissed the allegations in *Shell and BP in South Africa* – 'There is nothing new in the report,' the company commented.[19]

One month later the *Financial Times* reported that 'BP said that neither it nor its subsidiaries in Southern Africa supplied any oil to Rhodesia'.[20] The newspaper added that BP 'had resolutely complied with both the [UK] sanctions and South African law'. It was later discovered, however, that privately BP had admitted to the Government that its South African subsidiary was supplying some oil products to Freight Services. For on 4 March 1977, four days after publication of the Haslemere Group/Anti-Apartheid report, Montague Pennell, Deputy Chairman of BP, called at the Foreign Office for a meeting with Sir Anthony Duff, the Deputy Under-Secretary dealing with African affairs. According to BP's own account of the discussions, Sir Anthony Duff confirmed that the Government was 'aware of and shared BP's understanding of the situation affecting sale of products by BP Southern Africa (Pty) to Freight Services'.[21]

The Deputy Under-Secretary at the Foreign Office responsible for Africa therefore knew that BP *was* selling oil products to Freight Services for onward shipment to Rhodesia. Yet just a few weeks later a Foreign Office Minister told Parliament exactly the opposite. Minister of State Lord Goronwy-Roberts informed the House of Lords on 2 May 1977 that 'assurances of a substantial character from the highest level have been given us that no oil from British companies, including BP, finds its way directly or indirectly to Rhodesia'.[22]

Shortly after Lord Goronwy-Roberts' astonishing statement, Lord Brockway, a veteran anti-apartheid campaigner, wrote to seek clarification about the assurances. The Foreign Office Minister replied in further detail:

Senior representatives of the British oil companies have given assurances about the supply of oil to Rhodesia. They are to the effect that neither Shell nor BP nor companies in which they 'have an interest' are supplying oil to Rhodesia. Further, that they are not involved in any direct or indirect sale of oil products to Rhodesia except for the occasional and unavoidable 'little trickle' of certain base lubricants and bitumen which pass through local processors.[23]

Lord Brockway was being told a slightly different story from that given to the House of Lords: a 'little trickle' of oil products was now admitted to be supplied to Rhodesia by Shell and BP. In fact, however, Lord Goronwy-Roberts' 'trickle' probably represented most of Rhodesia's consumption of certain essential oil products. During BP's meeting with the Foreign Office in March 1977 the company had admitted that 'relatively small quantities' of lubricants were being supplied to Rhodesia.[24] But BP was probably using the words 'relatively small' to mean 'small in comparison with South Africa's total consumption of lubricants', rather than in comparison with Rhodesia's much smaller needs. Indeed, it is likely that the Shell/BP refinery at Durban was then supplying most of Rhodesia's requirements of lubricant base-stock processed at the Shell blending plant at Willowvale. Lord Goronwy-Roberts' description of the flow being a 'little trickle' therefore strayed from the truth. Lord Goronwy-Roberts also told Lord Brockway that this little trickle was 'unavoidable'. Again this later appeared to have been untrue: two months after the Minister's letter, BP South Africa actually cut off all sales to Freight Services.

Publicly at least, the British Government continued to maintain the fiction that Shell and BP were not involved in busting sanctions. Foreign Secretary Dr David Owen, wrote on 19 May 1977: 'We have repeatedly received high level assurances over the years from the oil companies that neither they nor their subsidiaries are supplying oil to Rhodesia.'[25] The truth, however, was rather different. On several occasions Shell and BP had privately admitted to the Government that their subsidiaries were supplying Rhodesia via Freight Services. Yet when the Haslemere Group/Anti-Apartheid report made this precise allegation, it was greeted by a flurry of denials both from the Government and the oil companies.

51

By this time Bernard and I had become quite accustomed to the stream of false information emanating from Shell Centre, BP's Britannic House, and the Foreign Office. Some of the statements were deliberately misleading; others, while strictly correct, were made to conceal reality – a form of distortion that Peter Kellner of *The Sunday Times* later christened 'petroleum truth'. But many of the oil company and Government denials were quite untrue. It is possible that some of the people who gave these comments were genuinely unaware that they were not telling the truth. But surely the press officers of Shell and BP and official Government spokesmen had a responsibility to find out what the situation was before they gave on-the-record statements? It must be concluded that a number of oil company executives, senior civil servants, and politicians were deliberately misleading the public or even lying in an attempt to cover up the sanctions scandal.

*   *   *   *

Behind the scenes, it was later discovered, publication of *Shell and BP in South Africa* had considerable impact. The British Government's majority shareholding in BP enabled the Treasury to nominate two Directors to the company's board to watch over the public interest. They also have the special right to veto any decision on certain specific matters – including foreign affairs. The supply of oil to Rhodesia, in contravention of a sanctions policy which had been initially proposed by the British Government itself, was exactly the sort of issue on which these two Directors should have focused their attention.

Tom Jackson, who was then one of the Government-appointed Directors, *did* take a strong line over the Rhodesian allegations. As leader of the Post Office Workers Union he had already acquired a reputation as a militant over Southern Africa issues, and just two months before he had led the controversial attempt to refuse to handle mail destined for South Africa as a part of an international week of action against the apartheid regime. Peter Hain, a well-known anti-apartheid activist, was then Jackson's research assistant, and he played an important role in encouraging his boss to take a decisive stand over Rhodesia. Jackson personally ordered fourteen copies of the Haslemere Group/

52

Anti-Apartheid report to distribute to members of the BP Board of Directors. At the Board meeting, three days after publication, the union leader asked about Rhodesia, and received some rather bland denials which left him far from satisfied.

Jackson then discussed the issue with his fellow Government Director, Lord Greenhill, who knew considerably more about the background of the sanctions affair. In his former job, as head of the Foreign Office, Lord Greenhill had learned of the Shell/BP swap arrangement with Total, and on two occasions, in 1973 and 1976, he had also been sent to Salisbury as a Government emissary for discussions with the Rhodesian regime. Lord Greenhill was therefore particularly well qualified to understand the significance of the allegations against the British oil companies. But Jackson soon found that his colleague had much less interest in this sensitive issue. It was therefore Jackson alone who made his way to the Foreign Office in the first week of March 1977. Although the trade union leader was anxious to deal with the situation discreetly, he was aware that he had one weapon left: if he remained unconvinced of the determination of BP and the Government to deal with the sanctions problem, it was open to him to step down from the BP Board. This, he realised, would cause a major political row, and force the issue into the open.

BP appeared to be living up to its advertising slogan in South Africa: 'No matter who you are – we like to keep you moving.' Jackson remained uneasy at the answers he was given by the Minister of State. 'Either the Government takes action,' he told Ted Rowlands bluntly, 'or I resign.'[26] The Government's façade of injured innocence was about to crumble.

NOTES

1 Mobil Rhodesia memorandum, *The Oil Conspiracy*, p. 41–2.
2 Letter from Richard van Niekerk (Mobil Rhodesia) to R. H. Maskew (Mobil South Africa), 2 September 1968, *The Oil Conspiracy*, p. 38.
3 Letter from Nicholas Gubb (Mobil South Africa) to W. J. Jackson (Mobil Rhodesia), *The Oil Conspiracy*, p. 16.

4 *Platt's Oilgram*, 8 July 1976.
5 *Sunday Times*, 11 July 1976.
6 Letter from Shell, 20 July 1976.
7 Letter from R. M. Goss (BP), 19 July 1976.
8 The Southern Rhodesia (United Nations Sanctions) (No. 2) Order 1968, section 5(1) (b & c).
9 UN Sanctions Committee, 9th annual report, p. 299.
10 UN Sanctions Committee, 9th annual report, p. 304.
11 UN Sanctions Committee, 10th annual report, p. 251.
12 Interview, 1 May 1979.
13 Written reply from Royal Dutch Petroleum to Kairos in response to a questionnaire from Kairos dated 13 June 1974.
14 *Shell and BP in South Africa*, pp. 30–1.
15 Hansard, House of Commons, 2 March 1977, col. 346–7.
16 Shell statement, 1 March 1977.
17 *Financial Times*, 2 March 1977.
18 *Guardian*, 13 May 1977.
19 *Guardian*, 4 March 1977.
20 *Financial Times*, 9 April 1977.
21 BP Submission to the Bingham Inquiry, 27 September 1977, p. 93.
22 Hansard, House of Lords, 2 May 1977, col. 877.
23 Letter from Lord Goronwy-Roberts to Lord Brockway, 24 May 1977.
24 BP Submission to the Bingham Inquiry, 27 September 1977, p. 93.
25 Letter from Dr David Owen to Tiny Rowland, 19 May 1977.
26 Johannesburg *Sunday Times*, 1 May 1977.

3   The Crumbling Cover-Up

Behind the scenes the British Government had come under mounting pressure during the early months of 1977 over the allegations being made against Shell and BP. One of the main forces behind this pressure was another multinational which had suffered severe financial losses through the activities of the oil companies: Lonrho. Lonrho, short for the London and Rhodesian Mining and Land Company, held the major shareholding in a pipeline that had been built to carry crude oil from the Mozambican port of Beira to the Rhodesian refinery at Umtali. The pipeline first came into operation in March 1965, just eight months before UDI; by the end of the year it had been closed as a result of the oil embargo, and shortly after, thanks to the Beira Patrol, all deliveries of crude oil supplies for Rhodesia were blocked. Since 1966 the closure of the Beira-Umtali pipeline has cost Lonrho nearly £100 million in lost revenue.

Tiny Rowland, the Chief Executive of Lonrho, began his crusade against the oil sanctions-busters early in 1976. His initial burst of anger followed the completion of a British Government investigation into Lonrho. The background to this inquiry went back three years to a boardroom row in 1973, which led to widespread allegations of business mal-practices. It was Lonrho's rather adventuresome style of doing business that Prime Minister Edward Heath was referring to when he coined the phrase 'the unacceptable face of capitalism'. The Department of Trade had subsequently set up an investigation into the activities of the company, and the results were finally published on 6 July 1976.

The most damaging section in the 660-page Department of Trade report was the long chapter on 'Lonrho's Activities in the Republic of South Africa and Rhodesia', in which accusations were made that Rowland and two other

Directors had been busting sanctions. These Directors, the report stated, had been 'more closely involved in matters relating to the financing and in consequence the development of the Inyati mine [in Rhodesia] than was consistent with the terms of UK sanctions legislation'.[1]

The London and Rhodesian Mining and Land Company was established in 1909. Rowland himself had gone out to Rhodesia in 1947, where he lived until UDI. But after joining Lonrho in 1961, he had realised that the economic future of the company must lie in the independent African states to the north. By the 1970s around half of the company's investments were in black Africa – an unusually high proportion for a British firm. Realising that revelations of sanctions-busting by Lonrho could prove a severe setback for the company's expanding interests north of the Zambezi River, Rowland decided to launch a counter-attack, claiming that Lonrho's activities were quite minor compared with the involvement of the oil companies in Rhodesian trade. Rowland, who was angry that the Department of Trade inquiry had been set up in the first place, was also aware that an exposure of the failure of oil sanctions would embarrass the British Government.

On 5 April 1976, after receiving an advance copy of the Department of Trade report, Rowland wrote to the Government with the charge that Shell and BP had provided over half of Rhodesia's oil since UDI. He recalled that on 22 May 1967 he had visited James Bottomley, a senior official at the Commonwealth Relations Office, to present documentary evidence of this trade. This included a list of chassis numbers of 300 railway tank-cars used to transport Shell and BP oil into Rhodesia. Bottomley, 'although he was at all times most courteous', apparently did nothing.[2]

Rowland then added a fascinating detail. As far back as 1967 Lonrho had considered embarking on legal action against the British Government over oil shipments to Rhodesia. Professor Raul Ventura, a Portuguese lawyer, had been requested to prepare a writ. But at this point Angus Ogilvy, a fellow Lonrho Director, raised problems over his connection with the Royal Family, and complained that his marriage to Princess Alexandra put him in a sensitive position. On the advice of Sir Michael Adeane, the Queen's Private Secretary, Ogilvy declared that he would have to resign from the Lonrho Board if the company launched legal proceedings against the Government. Lonrho's legal case was dropped to keep Ogilvy on the Board.

Rowland's letter went on to discuss the Beira Patrol, which had stopped the delivery of crude oil to Rhodesia, and had been introduced supposedly to show 'the steely determination of the British Government'.[3] But what was not given publicity, wrote Rowland, was 'the untrammelled and constant off-loading of refined petroleum products destined for Rhodesia'. His most astonishing claim was that 'over fifty per cent of these petroleum products were imported into Rhodesia by BP and Shell'. After these allegations were later released to the press, the oil companies' response was swift. 'This is absolute rubbish,' BP told the *Financial Times*; 'there is just no truth in it.'[4] The newspaper added that 'a Foreign Office spokesman said that they had looked into Mr Rowland's allegations and had not found anything to warrant reference to the investigating or prosecuting authorities'.

It was not until the beginning of 1977 that Rowland began in earnest his battle against the oil giants and the British Government. In the first few months of the year he sent the Department of Trade a stream of very long and detailed letters, all of them containing serious accusations against Shell and BP. But the Government – still hoping to cover up the scandal – virtually ignored his allegations. The correspondence between Lonrho and the Government soon settled into the following pattern: Rowland would write a very lengthy letter (sometimes running to over twenty pages); it would then take several weeks for the civil servants to send Rowland a formal two- or three-line acknowledgement; and then within hours of receiving this reply Rowland would fire off a further letter with yet more startling allegations. The process would then be repeated.

Rowland's correspondence provides a fascinating insight into the attitude of the British Government. One might have expected the authorities to welcome information which might lead to a stricter enforcement of sanctions, and the prosecution of firms that had broken the law. But the response to Rowland's information suggested that the Government was simply not interested. Indeed, it seemed positively anxious for his story *not* to surface.

On 7 January 1977 Rowland wrote a 12,000-word letter to Trade Secretary Edmund Dell, in which he made the startling claim that the oil companies, through their actions, had actually encouraged UDI. He went on to declare that Shell and BP began direct shipments to Rhodesia in December 1966, just a year after UDI, and that by 1967 the British-owned companies 'overtook everyone else'.[5] The oil embargo was reduced to tatters, Rowland pointed out, by 'a fatal disease – greed'.[6]

The Trade Secretary took over a month to tell Rowland that his letter had been 'forwarded to the appropriate authorities, who are undertaking further inquiries'.[7] Rowland replied the following day with another long letter making still further allegations against the oil companies. Enclosed were copies of documents from the Ministry of Trade in Salisbury which suggested that the oil companies had built up large stockpiles in Rhodesia at the time of UDI. Soon afterwards, he explained, these stocks played a vital role in cushioning the rebel regime against the initial impact of sanctions. Rowland added that he would be 'happy, under the cover of total secrecy, to fill the [Department of Trade] Inspectors' ears with damaging and juicy morsels which no man in his right mind would utter in an open court'.[8]

Why, Rowland asked, had there been no inquiry into sanctions-busting by Shell and BP, while the investigation into Lonrho had taken two years, and 'was caused by a boardroom split, not by any criminal act or by bankruptcy'?[9] Rowland concluded:

> It seems to me that if Lonrho was of some interest to your Department, then companies which directly maintain an illegal regime in power for over eleven years, keep its military aircraft and helicopters flying, and literally fuel Rhodesia's entire economy, offer a much wider field in which to deploy your Inspectors. Your Department has had the powers and the information for almost a dozen years – why haven't they been used?

Not surprisingly the oil industry quickly came to regard Rowland as one of its bitterest enemies. A BP spokesman later admitted to a journalist that 'we have a wax model of him, and we are sticking pins in it'.[10]

Meanwhile events were moving quickly. Rowland had learned that the Haslemere Group/Anti-Apartheid report on *Shell and BP in South Africa* was due to be published on 1 March 1977, and the day before, he shot off a further letter to the Trade Secretary. This time Rowland gave astonishingly detailed statistical data on the quantities of oil supplied by Shell and BP since UDI. In 1973, when imports reached their height, he revealed that the British-owned companies were providing fifty-six per cent of the 4,300,000 barrels of oil which were railed into Rhodesia from Lourenço Marques. Rowland went on to suggest that there had been some form of secret agreement between the oil companies operating in Rhodesia to retain almost exactly the same share of the market as each had held before UDI. This was not, he explained, pure coincidence. The oil companies had clearly been anxious to keep competitors out of the Rhodesian market.

On the day that the Shell and BP report was released, Rowland personally rang the Anti-Apartheid office to order copies. He spent an hour on the phone, talking of the evidence that he had gathered against the oil companies. Since Ian Smith had recently agreed to the principle of majority rule, Rowland explained, it had become much easier to get people to talk. 'It was like World War Two: in 1942 no German officer would admit that there were concentration camps, but by 1945 they were queuing up to tell everything.' Rowland finished the long phone conversation by revealing the astonishing news that Lonrho planned to sue Shell and BP for 'many millions of pounds'. Bernard Rivers and I were amazed to hear of the evidence which Rowland appeared to have obtained. Until this point none of the damaging allegations which he had made privately in his letters to the Department of Trade had been published.

Lonrho's Directors met on 7 March 1977 to consider the most controversial issue that they had faced since the boardroom row that had torn the company apart four years earlier. Rowland, backed by several Directors, pushed for a decision to proceed with legal action against the oil companies for continuing to supply Rhodesia after the closure of the Beira-Umtali pipeline. But Lord Duncan-Sandys, Lonrho's Chairman, was hesitant. As a Conservative MP and former Commonwealth Secretary, he had personally negotiated the

1961 Rhodesian Constitution, and his views on the situation therefore held great weight. The Lonrho Board decided to delay legal action for a little longer.

The correspondence game continued. On 14 March the Trade Secretary sent a brief reply stating that Rowland's earlier letter had been forwarded to the Foreign Office. Copies of two questions in the House of Commons were attached. On 30 March Rowland wrote back in terms which were distinctly cool:

> I was disappointed by your kind letter of the 14th, with a Hansard extract. As part of my business, I follow any relevant debate or question in the House of Commons . . . I felt that it was a bad swap to receive three pages of Hansard in return for key documents.[11]

The Lonrho chief was becoming increasingly frustrated at the response he was receiving from the British Government, but this only strengthened his belief that the company should intervene directly by initiating legal proceedings against the oil companies.

*　*　*　*

Rowland's main source of ammunition in his battle against the oil companies was a fascinating Portuguese figure, Jorge Jardim, who had his own reasons for wanting to unmask the hypocrisy of the sanctions-busters. As a Portuguese business-man, living in Mozambique, he had run the Sonarep refinery at Lourenço Marques in the mid-1960s and had therefore played a key role during the crucial post-UDI period in helping Rhodesia import its oil supplies. But Jardim's power went beyond his business interests. He was also a man of immense political influence – mainly because of his close personal ties with the former Portuguese dictator, Dr Antonio Salazar. It was often said that he was even more powerful in Mozambique than the Governor General.

Jardim was almost fanatically sympathetic to the memory of the late Dr Salazar. After the Portuguese coup in 1974, which had toppled Dr Salazar's successor, Jardim had started to write a series of books on developments in former

Portuguese-ruled Africa, one of the most important topics he sought to expose being the long-established myth that Dr Salazar was to blame for Rhodesian UDI. 'It was time to finish with the farcical explanation that it was Portugal which was responsible for the failure of sanctions', he later recalled. 'It was the international oil companies themselves which drove a wedge through the oil embargo.'[12]

Jardim began work on a book exposing the sanctions scandal during 1976. Because of his influential position in Mozambique, in both business and administrative circles, he had always had access to an immense amount of information on Rhodesian trade. He had also been very careful to collect as much documentation as possible, and this huge library of papers provided the basis for his account. By the end of 1976 he had already completed the first draft of his book, when a Portuguese business colleague told him that he must talk with Rowland. Jardim, who had known Rowland for many years, asked why, and was told that Lonrho had begun to gather evidence for use in a legal case against the oil companies.

Jardim immediately phoned Rowland, and the two men soon met. 'When I told Rowland I was writing the book', Jardim later recalled, 'he immediately jumped up in excitement'.[13] Quickly the two men struck a bargain. Jardim would provide information and documentation for use in Lonrho's law case, and in return Rowland would hand over his material for Jardim's book. So it was that many of the allegations which Rowland made in his series of letters to the British Government during the early months of 1977 originated from material provided by Jardim. At this time, however, the Foreign Office was probably unaware of the identity of Rowland's major informant.

*   *   *   *

Meanwhile Rowland had found African allies in his battle against the oil giants. During the summer of 1976 it looked as if a peaceful settlement might be reached in Rhodesia. US Secretary of State Dr Henry Kissinger's diplomatic shuttle through the capitals of Southern Africa had created the basis of an agreement. The African nationalist groups and the Smith regime agreed to talk, and on 28 October 1976 both

61

sides faced each other across the table at Geneva. But as the weeks dragged by, it became increasingly clear that a settlement would never emerge out of the Geneva Conference.

On one occasion, during a lull in the formal negotiations, private conversation among some of the nationalists turned to sanctions. How was it, they asked themselves once more, that Smith had been able to remain in power for over a decade? And, in particular, how did Rhodesia import its vital oil requirements? One of those present was Justice Robert Hayfron-Benjamin, a distinguished Ghanaian High Court judge who had been seconded as a constitutional advisor to the nationalist delegations. When the discussions came round to sanctions, he remembered that an old acquaintance of his had always had a special interest in Rhodesia's oil. Justice Hayfron-Benjamin immediately telephoned Rowland in London. The two men quickly agreed to meet, and the Lonrho chief told the Ghanaian judge about the evidence that he had been gathering for his law suit against the oil companies. As the conversation went on, the idea developed that it might also be possible for the Rhodesian nationalists to launch their own legal case.

For the next few months, Justice Hayfron-Benjamin spent much of his time at Lonrho's Cheapside headquarters in London. With considerable assistance from Lonrho, the Ghanaian judge began to assemble the Rhodesian nationalists' case against the oil companies. Rowland, who was confident that majority rule would soon come to Rhodesia, believed that support for the nationalists was vital for Lonrho's substantial investments in the country, and regarded the goodwill of Rhodesia's future black leaders as a form of insurance.

On 20 December 1976 Justice Hayfron-Benjamin sent formal letters to Shell, BP, Mobil, Caltex, and Total on behalf of 'all the African nationalist delegations' at the Geneva Conference. 'Overwhelming evidence' suggested that the companies had conspired 'to maintain the illegal Government in power'.[14] His instructions, he continued, were 'to prosecute all the claims that the Africans have against your group's involvement in the massive conspiracy to violate sanctions'. Although strongly worded, the letter gave no substantive evidence that the oil companies had been involved in sanctions-busting.

It was at the end of December 1976 that Justice Hayfron-Benjamin first saw *The Oil Conspiracy* report, and read Mobil's subsequent defence before the US Senate. This gave the Ghanaian judge a new idea. If it was possible to produce additional evidence against Mobil, he realised, then the Rhodesian nationalists might be able to get the Senate to reopen hearings. By this time, however, he was no longer acting on behalf of all the Rhodesian nationalists – as he had earlier claimed – but only for Bishop Abel Muzorewa's United African National Council (UANC).

On 9 March 1977 Justice Hayfron-Benjamin wrote to Senator John Sparkman, Chairman of the Senate Foreign Relations Committee, to inform him that the UANC 'would welcome the opportunity to place before your subcommittee on African affairs, material which goes to prove that two American oil companies, namely Mobil and Caltex, have been involved with other oil companies, namely Shell, BP and Total . . . in a conspiracy to break sanctions.' Accompanying the letter was a 12-page memorandum containing detailed allegations against the five oil companies. The oil companies did not appear unduly concerned by the UANC's attempt to reopen Senate hearings: Mobil confidently believed that its testimony to the Senate the previous September had disposed of the charges once and for all, and BP, when asked to comment on the UANC memorandum, would only 'reiterate what we have said before – BP adheres to all the UK Government's laws and does not contravene its policy towards South Africa and Rhodesia'.[15]

But the oil companies were acutely aware that the dangers they faced came not so much from the Rhodesian nationalists, but from independent Africa. Justice Hayfron-Benjamin's letter to the companies had contained a worrying sting in the tail – the Rhodesian nationalists, he said, were 'contemplating legal action in the Federal Courts of . . . Nigeria.'[16] This caused considerable concern to the oil companies, who had massive investments in Nigeria worth many hundreds of millions of pounds. Indeed only a few months earlier the Chairman of Shell had specifically warned the Foreign Office that 'any public escalation of this sanctions issue can only make an already difficult situation more difficult in relation to our . . . interests in black Africa (especially, perhaps, Nigeria)' (B/A 284).

The Nigerian Government had been making increasingly militant statements about using its economic muscle to retaliate against companies deeply involved in sustaining minority rule in Southern Africa. The oil companies' fears therefore increased a few weeks after Justice Hayfron-Benjamin's letter, when Nigerian Foreign Minister Brigadier Joseph Garba told them that they would have to choose between doing business with black and white Africa. 'If Western oil monopolies continue to ignore the UN decisions against apartheid', he warned, 'we will take appropriate measures'.[17] Rumours quickly circulated in business circles that the assets of the oil companies in Nigeria might be frozen as a reprisal against their involvement in Rhodesian trade. But the oil companies retained one strong card: Nigeria remained dependent on their technological skills. The oil companies' interests in Nigeria therefore appear to have emerged almost unscathed from the sanctions scandal.

Clearly however, Nigeria was not the only African state concerned about the sanctions issue. Justice Hayfron-Benjamin also had the backing of his own government in Ghana, and produced a brief report for them, based on material provided by Lonhro, which gave detailed statistical data on the flow of oil to Rhodesia. In the first half of 1967, for instance, the Southern African subsidiaries of Shell and BP despatched 2,346 railway tank-cars from Lourenço Marques into Rhodesia containing 574,000 barrels of oil products; Mobil (267,000 barrels), Caltex (285,000 barrels), and Total (286,000 barrels) provided most of the remainder.

Justice Hayfron-Benjamin's report formed the basis for an official Ghanaian statement which accused the oil companies of 'criminal responsibility in supplying Rhodesia with her oil requirements over the past eleven years . . . and thus helping in prolonging the existence of an illegal regime'. The Ghanaian statement was important in being the first published account of the exact quantities of oil supplied by the various oil companies. But being issued on Christmas Eve 1976, its unfortunate timing meant that it received only local press coverage.

*   *   *   *

The strongest call for a tightening of oil sanctions came from Zambia, Rhodesia's northern neighbour, which had suffered enormously from UDI. The Zambian Government strongly opposed the Smith regime, and at the time of UDI President Kaunda had called on Britain to use force to topple the rebel regime. Sanctions, the Zambians predicted, would prove ineffective. But in spite of this the Zambians went ahead and cut economic links with their neighbours, at tremendous cost to themselves. The United Nations has estimated that the additional costs incurred by Zambia as a result of UDI have now exceeded £500 million.

In December 1976 President Kaunda was given a copy of *The Oil Conspiracy* report, and was furious to discover that his earlier suspicions that the international oil companies were fuelling Rhodesia had indeed been correct. The escalating war in Rhodesia, he therefore believed, was partly the responsibility of the oil companies; so, too, were the eleven years of UDI which had caused such enormous financial hardship to his own country.

Kaunda spoke out publicly against the involvement of the oil companies on 29 January 1977. He opened the Liberation Committee of the Organisation of African Unity by accusing the oil companies of responsibility 'for the deaths of thousands in Zimbabwe (Rhodesia) and for Smith's attacks against independent African states'. The oil companies must 'choose between Ian Smith and cooperation with Africa'. They could no longer 'run with the hare and hunt with the hounds'.

Two days later Kaunda had his first opportunity to express his anger to a Western emissary. Ivor Richard, Britain's UN Representative and Chairman of the abortive Geneva Conference, was in Lusaka for discussions over Rhodesia. He was invited to State House for the evening. But to Richard's astonishment, he was greeted by a very angry President, full of rage at Britain's hypocrisy over Rhodesia in general, and over oil supplies in particular.

One Zambian Minister later commented that the President had 'drenched' the unfortunate Richard. Half an hour later the stunned British envoy emerged from State House, having virtually been thrown out of the President's office. With him he clutched a letter which Kaunda had asked him to pass on to the British Prime Minister, James Callaghan. The Zambian President was virtually calling for a diplomatic break with Britain.

Callaghan replied to the angry Zambian President's letter on 11 February 1977. 'Dear Kenneth,' he wrote by hand:

> I am shocked and astonished that you should not only say but apparently believe that I have been cheating you for years, both as Chancellor of the Exchequer and as Foreign Secretary over the matter of oil sanctions, on the grounds that I must have known what 'Shell had been doing'. On the contrary, during my two years as Foreign and Commonwealth Secretary I went to great pains to see that sanctions were maintained and indeed tightened up ...
>
> You raised with Ivor Richard the question of our personal relations from now on. I will not go into that here because I find it too painful, having been a faithful friend of yours and of African independence for thirty years or more. Deeply though I regret the break in our personal relations, what is perhaps even more important is that our two countries should understand one another and, Kenneth, you must understand, if I have evidence about sanctions-breaking, then that evidence will be followed up and, if it is possible to bring the culprits to book, then that will happen ...

This personal commitment by Callaghan was to have an important impact on British policy some eighteen months later, but in the short term his reply did little to mollify the Zambian leader. It was at this point that Kaunda decided to take up the suggestion that had been put to him a few weeks earlier, that Zambia should take the oil companies to court. The idea had come from Rowland, a frequent visitor to Lusaka and one of the few businessmen to have close links with Kaunda. (On one occasion in 1977 Kaunda described Rowland as 'one of the six good capitalist friends that I have', leading to endless speculations in Lusaka as to the identity of the other five!)[18] Rowland had realised that if he was able to assist the Zambian Government in preparing a law case against the oil companies by providing evidence and legal assistance, he would then increase his goodwill in an important black African state. He therefore pledged his full support to Kaunda's government in pursuing the oil companies.

The Zambian Ministry of Legal Affairs quickly began to assemble evidence that the oil companies had sustained the rebel regime, and by 23 March 1977 the first draft of Zambia's Statement of Claim had been completed. One week later Kaunda was quoted in an interview in *The Times* as saying that his Government planned to take the oil companies to court over sanctions-busting. But the paper's Southern African correspondent apparently did not understand the significance of the remark, since two sentences about the case were buried in a longer article.[19] Four days later The *Observer* announced that Zambia was 'actively considering' legal action against five Western companies for 'oiling Rhodesia's war machine' in defiance of UN sanctions. The newspaper added that 'a spokesman for Shell International in London yesterday denied that the company was in any way evading sanctions'.[20] The following day one British oil company executive complained that there was 'a growing and carefully orchestrated campaign against us by certain African leaders'.[21]

By the spring of 1977 the British Government and the oil companies were under increasing pressure to take action over allegations of sanctions-busting. My report on *Shell and BP in South Africa*, published on 1 March, had led to public discussion on the issue. But at this time the pressures from Lonrho, the Rhodesian nationalists, and the Zambian Government had been almost entirely behind the scenes. Nevertheless, the British Government realised that it was probably only a matter of weeks before the different accusations came together and the issue could well blow up as a major political scandal.

*　　*　　*　　*

One of the fascinating aspects of the breaking of the sanctions story was the diverse array of people involved: Rowland, the unconventional business figure, had joined forces with the Rhodesian nationalists and the Zambian Government, who clearly had interests rather different from those of Lonrho; Jardim, politically on the far right, was trying to clear the name of a former Portuguese dictator – and

Bernard Rivers and I were two young radical researchers who smelt a scandal. Together we made a most curious collection of investigators.

It was one of the strange twists of history that a full decade elapsed after the introduction of the oil embargo before any serious effort was made to uncover the sanctions story, and that when this did occur, in the mid 1970s, a number of people began to work on the subject quite independently. It was not until 1976, after he had finished the draft of his book, that Jardim discovered that Rowland was gathering evidence for a law case against the oil companies; and Bernard and I had no idea that other people were also pursuing the sanctions trail until early 1977. So why was it that interest in the oil embargo suddenly emerged from different quarters at almost exactly the same moment? One simple answer is the coup in Lisbon, and the subsequent independence of Mozambique, which opened up new opportunities for research. But there also seemed to be a large element of chance at work in bringing a number of people in different places to pursue the same story at the same time. Certainly the more information we all obtained, the more we all realised the full extent of the scandal which we were uncovering.

The fact that by early 1977 allegations were being hurled at the British Government from such diverse sources no doubt played an important part in forcing the Foreign Office to act; and the Government was particularly worried by Tom Jackson's resignation threat. But President Kaunda's anger was the last straw. On Easter Sunday Foreign Secretary Dr David Owen was due to fly to Lusaka on the first stop of a fact-finding safari to see whether the deadlocked Geneva Conference could be salvaged. Zambia's support for any new initiative over Rhodesia was essential, and the Foreign Secretary realised that something had to be done to ensure that he received a better reception than Ivor Richard had had in Lusaka two months earlier.

On Good Friday, 8 April 1977 Dr Owen therefore announced the establishment of an official inquiry. It was a strange day to announce such an initiative, but it was vital to win back Zambian goodwill. The Foreign Secretary explained that under the terms of the Sanctions Order he was empowered to set up an investigation of possible breaches of the law. An inquiry was therefore to be set up with four purposes:

    (a)   To establish the facts concerning the operations whereby supplies of petroleum and petroleum products have reached Rhodesia since 17 December 1965;

    (b)   To establish the extent, if any, to which persons and companies within the scope of the Sanctions Orders have played any part in such operations;

    (c)   To obtain evidence and information for the purpose of securing compliance with or detecting evasion of the Southern Rhodesia (United Nations Sanctions) (No. 2) Order 1968 (the Sanctions Order);

    (d)   To obtain evidence of the commission of any offences against the Sanctions Order which may be disclosed.

More than eleven years after UDI the British Government was to *begin* an investigation into how oil was reaching the rebel regime.

NOTES

1 *Inspectors' Report on Lonrho Limited*, (HMSO, 1976), p. 647.
2 Letter from Rowland to H. C. Gill (Department of Trade), p. 3.
3 Letter from Rowland to H. C. Gill (Department of Trade), p. 3.
4 *Financial Times*, 7 July 1977.
5 Letter from Rowland to Edmund Dell (Department of Trade), 7 January 1977, p. 21.
6 Letter from Rowland to Edmund Dell (Department of Trade), 7 January 1977, p. 11.
7 Letter from Edmund Dell (Department of Trade) to Rowland, 10 February 1977.
8 Letter from Rowland to Edmund Dell (Department of Trade), 11 February 1977, p. 4.
9 Letter from Rowland to Edmund Dell (Department of Trade), 11 February 1977, p. 5.
10 *Business Week*, 23 September 1978.
11 Letter from Rowland to Edmund Dell (Department of Trade), 30 March 1977, p. 1.
12 Interview, 10 February 1979.

13 Interview, 10 February 1979.
14 Letter from Justice Robert Hayfron-Benjamin to Mobil, 20 December 1976.
15 Letter from John Collins (BP) to Peter Kellner (*Sunday Times*), 18 March 1977.
16 Letter from Justice Robert Hayfron-Benjamin to Mobil, 20 December 1976.
17 *Jornal de Angola*, 30 January 1977.
18 *Financial Times*, 26 October 1977.
19 *The Times*, 1 April 1977.
20 *Observer*, 3 April 1977.
21 *Guardian*, 4 April 1977.

4   The Smoking Gun

Thomas Bingham QC was spending Easter holiday weekend
at his cottage in the Welsh mountains when he received a
phone call from the Foreign Secretary asking him to head
an investigation into oil sanctions. He had a busy legal practice, and some cases would have to be completed first, but
he told Dr Owen that he would be happy to take the job.
Neither Dr Owen nor Bingham probably imagined on that
Good Friday in 1977 what an extraordinary story the official
inquiry would tell.

The Foreign Office officially appointed Thomas Bingham
on 10 May 1977. But it was not until 19 July that Marius
Gray, an accountant, joined the inquiry, and even at this late
stage a part-time oil expert had yet to be appointed. It certainly seemed a very small team for an urgent investigation
which would cover the activities of Britain's two largest
companies in four principal countries (Britain, South Africa,
Mozambique, and Rhodesia) over a period of nearly twelve
years. To begin with the inquiry looked as if it would take a
long time to complete – but there were other drawbacks: if a
Rhodesian settlement was reached, and the public lost
interest in sanctions, then there was nothing to stop the
report being quietly buried. And in the meantime the
Government and the oil companies could refuse to answer
any questions about sanctions-busting on the grounds that
the matter was already under investigation. Was it therefore
possible that the official 'inquiry' was merely a new stage in
the cover-up?

The hearings of the official inquiry were to take place in
private. Even if Bingham did manage to get at the truth,
there was no guarantee that the story would ever emerge,
since the Foreign Secretary was not obliged to publish his
report, and to do so would require the formal consent of all

71

the witnesses involved. This might mean that oil companies, which were to provide much of Bingham's evidence, might effectively hold a veto power over publication. It was also generally assumed that even after Bingham had handed in his report, it would take many months for the Foreign Secretary to 'study' its findings and decide whether to authorise publication. It was hardly surprising that some observers argued that Dr Owen had opted for the time-honoured expedient of an investigation in an attempt to defuse mounting pressure for real action.

Both Shell and BP immediately announced that they would offer their full cooperation to the inquiry. The Foreign Secretary had the authority under the Sanctions Order to compel witnesses to give evidence, and the oil companies therefore knew that they would have to assist the investigation. But in the end Shell and BP gave evidence willingly without being subpoenaed to appear before the investigation. Initially there were some disturbing signs that the oil companies were attempting to prejudge the inquiry. On the day that Dr Owen set up the investigation a BP spokesman told the press, 'we have never broken sanctions – United Nations, British or any other.'[1]

There were also fears that the oil companies might try to hide part of the truth. On 23 May 1977 President Kaunda announced that the oil companies had just held a conference in Zurich 'at which they decided to destroy all the important papers so that we would not have evidence with which to present our case to court'.[2] A Shell spokesman immediately dismissed this allegation: 'We are not aware of any meeting, secret or otherwise, between Western oil companies.'[3]

Later Bernard Rivers spoke with Dumisani Kumalo, a black executive from Total South Africa. He claimed that shortly before the Swiss conference, an industry meeting of all the major oil companies had been held in South Africa. In preparation Kumalo had been asked by his boss William Parsons to collect copies of all documents relating to Rhodesia, and to assemble them together in a bulky file. These included the relevant pages of a computer print-out giving monthly statistical data on Total's sales to Rhodesia. Kumalo never again saw these papers. Immediately after the industry meeting in South Africa he heard that the Total General Manager, Alphonso Hough, had attended an oil company meeting in Switzerland to discuss Rhodesian trade.

Despite Shell's claim that the Swiss conference never took place, Bingham confirmed that an important meeting had been held in Zurich in May 1977 (B12.8). Among those present were the General Manager of Shell Rhodesia (J. L. Whitehead) and representatives from Shell in London. Shell told Bingham that the meeting had been called to discuss terms for retaining Whitehead's services after the normal retirement date. But it was not clear why these discussions took place in Switzerland, of all places, and it is quite possible that other topics were also raised, possibly more informally. The fact that Bingham has confirmed the existence of the Zurich meeting and that Kumalo has independently claimed that Total executives were present adds credence to fears that some of the oil companies may have met to consider destroying documentation relating to Rhodesia.

13

Several informants whom I interviewed in Portugal also told me that Shell documents relating to Freight Services had been destroyed in Lourenço Marques at the time of the closure of the Mozambique-Rhodesia border. Several of these contacts were subsequently interviewed by Granada Television. One former employee of Shell Mozambique explained how he had burnt documents referring to supplies to Freight Services. His instructions, he claimed, came 'from the top'.[4] There were therefore disturbing signs that attempts had been made to destroy incriminating evidence held in the oil companies' files in Lourenço Marques.

Initially, Bernard and I were therefore far from convinced that the Bingham Inquiry would ever uncover the full story of the sanctions scandal. But we immediately agreed that it was vital to continue pursuing the story; firstly, if the Bingham Report was not published, it would be important for independent evidence to be reported in the press; secondly we realised that releasing fresh information while Bingham was carrying out his own investigation might arouse further public interest in the issue, and force the Government to proceed with speedy publication.

*   *   *   *

We began to assemble all the material we had accumulated in our files, and were quite surprised at the volume of the evidence we had already acquired. The material, we agreed, had to be placed before the official inquiry. Bernard and I therefore prepared a 10,000-word submission which was presented by the Haslemere Group and the Anti-Apartheid Movement on 25 April 1977.[5]

Initially we had decided to work together only for the ten hectic days it took to prepare the submission to the inquiry. In the end, however, we pursued the oil story together for eighteen months. Quickly we discovered which aspects of the work we could each do most effectively. Bernard was a first-class investigator when he came across an important new lead, and would put tremendous energy and imagination into pursuing an informant and ferreting out the facts. But after his initial enthusiasm for a particular contact had waned, he tended to lose interest and had to put up with my nagging him to follow up the leads that he had already developed. Since I found it easier to do the writing, I ended up with most of the journalistic work on the story.

We divided the world. Bernard took on the United States, where he was based much of the time, and pursued information relating to Mobil and Caltex, at the same time building up good contacts at the United Nations. I was a frequent visitor to Holland and France, following up various leads on Royal Dutch Petroleum and Total, and also went to Lisbon several times to interview former employees of Shell Mozambique who had recently returned to Portugal. This later proved particularly important in the light of the key role played by the joint Shell/BP company in Mozambique.

In the end we built up files on over two hundred oil executives, ranging from the Chairman of Shell to the humble worker at Lourenço Marques who physically loaded Shell consignments for Rhodesia. For each of these oilmen we tried to obtain details of posts held, photographs, addresses, extracts from various *Who's Whos*, press clippings, letters that they had sent or received, and other documents in which they were mentioned. When we eventually tracked the people down, it was often strange to meet them in the flesh after spending so long building up their character on paper.

The high costs of mounting an international investigation made our research extremely expensive. Harold Evans, editor of The *Sunday Times*, once asked us why we had not come to him many months before with our story and claimed that he would have been happy to have financed our investigations. Yet when we first began pursuing the sanctions trail it was extremely difficult to awaken any interest from the press. We would have been lucky even to have set foot inside an editor's office, let alone be taken on to cover the story. We therefore had to obtain financial backing to supplement our meagre journalistic earnings. Bernard and I worked as part-time consultants to the Commonwealth Secretariat and the United Nations on the oil sanctions issue. In addition the Haslemere Group in London and the American Committee on Africa in New York subsidised our research with funds obtained from church groups and anti-apartheid organisations. With their help Bernard and I were able to work full-time on the sanctions story.

*   *   *   *

It quickly emerged that the Zambian Government was as cynical about the British Government investigation as we were. President Kaunda was convinced that the inquiry was yet another way of delaying real action to tighten the embargo, and the apparent slowness with which the investigation got under way only seemed to confirm this impression. Why, Kaunda asked, had the British not yet discovered the facts? 'My little nation with its limited means, has been able to learn so much about how oil is reaching the Smith regime and how UN sanctions are being broken, that it is inconceivable to me that British intelligence services should not have known all about it.'[6] Zambia therefore decided to boycott the British inquiry, and a few months later the Zambian Government rejected a request by Bingham to visit the country officially.

Zambia decided to take its own initiative and went ahead with a law suit against the oil companies. On 2 July 1977 the Zambian Attorney General, Mainza Chona, wrote to the oil companies to inform them that legal proceedings would be initiated. He alleged that their shipments to Rhodesia had contravened a 1962 agreement between the oil companies and the Central African Federation (which included both Zambia and pre-independence Rhodesia). 'The necessary and direct result of your actions has been to cause and continue to cause considerable financial loss to Zambia, putting

75

her economy at very serious risk. The loss and damage, which is continuing, is at present conservatively estimated at not less than $1,000m.'

Work proceeded on the Zambian legal case, and on 24 August 1977 the writs were filed at the High Court in Lusaka. The accompanying Statement of Claim, which outlined the case against the oil companies, ended with a claim for damages that had grown to a total of £4,000 million. This made the case one of the largest in legal history. But Zambian officials stressed that the Government's motivation in bringing the action was not economic, but political. Kaunda's aim was to expose the sanctions scandal – and legal action, he believed, was an effective way to do it. Unfortunately the Rhodesian nationalists' attempt to take legal action against the oil companies foundered on the growing tensions between the nationalist groups and their external supporters. At the Geneva conference in late 1976, when the idea of going to the courts had first arisen, the various nationalists had been more willing to cooperate with each other. Justice Hayfron-Benjamin, who provided much of the initial impetus behind the scheme, was then acting on behalf of all the Rhodesian nationalists. But by the end of 1976, as differences between the Rhodesian groups hardened, he himself ended up working solely for Bishop Muzorewa's UANC.

'When we were ready to sue the companies involved,' Bishop Muzorewa later recalled, 'some people felt that if we did this credit would go to the UANC, and so they snatched it [the evidence] from our hands.'[7]

The fact was that Rowland, who supported the Rhodesian nationalists in order to protect his financial interests in the country, had by this time decided to give his backing to Joshua Nkomo's Zimbabwe African People's Union (ZAPU), one of the two groups in the Patriotic Front. In addition Rowland continued to provide some assistance to the Reverend Ndabaningi Sithole's internal wing of the Zimbabwe African National Union (ZANU). Neither ZAPU nor ZANU, however, showed much interest in pursuing legal action against the oil companies. The deteriorating relations between Lonrho and Bishop Muzorewa's UANC therefore effectively ended the attempt to sue the oil companies.

But the Rhodesian nationalists still continued to declare

76

publicly that oil companies were playing an important role in sustaining the Smith regime. Robert Mugabe, co-leader of the Patriotic Front, pointed out that 'the West talks of wanting peace – but in reality props up the minority regime through sales of oil'.[8] The Bingham Inquiry, the nationalists believed, was just a transparent delaying tactic. Joshua Nkomo echoed the view of many Rhodesian nationalists when he blurted out at a press conference: 'Inquiry! Inquiry into what? This is the British way of tackling things. They know what is happening.'[9]

Rowland was equally cynical about the official investigation. On 26 May 1977 the Lonrho chief wrote to Dr Owen to inform him that he did not intend to cooperate with the inquiry:

> It will be months and months before Mr Bingham has mastered his material, and the whole idea is handicapped from the start. One man, sitting *in camera*, attempting to summon reluctant foreign witnesses who have no obligation to appear, and expecting them to disgorge information in a way that can only damage their own careers – it's a non-starter, particularly as I know that many of the witnesses will be intimidated, either by fear of losing their livelihoods or by more direct threats.[10]

Rowland went on to recount how he personally was being intimidated. The local Director of Lonrho in South Africa had recently been summoned to the Prime Minister's office at Cape Town. There he was told by the Permanent Secretary of the Foreign Ministry: 'If Rowland continues to pursue the case against the oil companies, we will pull the trigger . . . and prosecute Lonrho'.[11] Rowland admitted to Dr Owen that the South African threat looked 'ridiculous and melodramatic', but he assured him nonetheless that the South African Government's hostility towards those who were exposing the sanctions scandal was genuine – and extreme.

Lonrho had undoubtedly played a key role in forcing the British Government to establish an official investigation into Shell and BP. Yet Rowland quickly emerged as one of the most outspoken critics of the inquiry, and as it turned out, he was the only witness who actually had to be subpœnaed to give evidence. On 21 November 1977 Bingham wrote to Rowland to say that he had 'a public duty to perform', and that he could not allow this to go by default. Accompanying the letter was a formal subpoena demanding that Lonrho provide seven different categories of documents in his possession relating to oil sales to Rhodesia. Rowland immediately complied with the subpoena, and the Lonrho chief ended up as one of the Bingham Inquiry's most valuable witnesses.

Meanwhile Lonrho went ahead in pursuing legal action against the oil companies. A decision to proceed with the law case was taken by the Lonrho Board just a few days after the establishment of the Bingham Inquiry. On 26 April 1977 Lonrho's solicitors, Cameron Kemm Nordon, sent formal letters to the oil companies claiming that since UDI they had 'wrongfully and unlawfully conspired' with Freight Services, Genta and the Smith regime 'to supply oil to Rhodesia'.

The next stage in Lonrho's legal battle was to issue the writs. On 31 May 1977 Lonrho filed a writ at the High Court in London against no fewer than thirty-three companies, the action also being directed at a number of subsidiaries of the five international oil companies. The writ was accompanied by a 55-page Statement of Claim which outlined the basis of Lonrho's case, pointing out that the oil companies, prior to UDI, had signed an agreement to use the Beira-Umtali pipeline – and no other route – to supply their Rhodesian subsidiaries. By later failing to prevent their South African subsidiaries from breaking sanctions, and by sending oil into Rhodesia by road and rail, the companies were therefore acting in breach of contract. Lonrho also claimed that by helping to ensure the survival of the illegal regime, the oil companies were partly responsible for the fact that the pipeline had remained closed for over a decade, with such a huge resulting loss of revenue to Lonrho.

Even after the Bingham Inquiry had been set up there continued to be widespread misgivings that the full sanctions story would not emerge, since even if Bingham managed to discover the facts about the involvement of British oil companies in sanctions-busting, the Government had no obligation to publish the report. These fears were hardly allayed when Dr Owen continued to answer Parliamentary questions about the publication of the Bingham Report by saying that he would first have to make a thorough study of the results of the inquiry.

At a press conference in June 1977 I asked Callaghan when the Bingham Report might be completed, and was told: 'I imagine that you know the difficulties that exist there under the South African Official Secrets Act . . . '[12] He went on to explain that in South Africa it was an offence to pass on information relating to oil sales. 'We are operating against that handicap at the moment.' Three days later the *Guardian* suggested that the Bingham Inquiry might not lead very far. The newspaper reported that 'the Foreign Office is already preparing the ground for a disappointing outcome with emphasis on the difficulties which the South African Official Secrets Act may place in obtaining the necessary commercial information from local subsidiaries'.[13] It appeared that the British Government might well be adapting the dubious 'Mobil defence'.

* * * *

Ultimately, however, it was the leaking of two key sets of documents which ensured that the Bingham Inquiry shattered the thirteen-year cover-up. These secret papers represented the 'Smoking Gun' which identified the villains behind the sanctions scandal. The most incriminating material was contained in a thin manilla foolscap folder with a small white label on the cover. On it had been typed the words 'Sandford File' and inside were 41 sheets held together by a metal clip. These papers had been in the possession of Arthur Sandford, a key London-based BP executive dealing with Southern Africa.

Little has been published on how these papers were leaked from BP. But although the identity of the original BP employee who leaked the papers is still unknown, it is now possible to trace the route through which they surfaced. The mysterious 'whistle-blower of Britannic House' as he was later named, had become extremely angry at the company's duplicity over sanctions-busting, and is understood to have given the bundle of papers to a contact; in July 1977 this intermediary passed the file on to Gerry Gable, a researcher with London Weekend Television. Gable then showed the documents to London Weekend and during the autumn of 1977 there was some talk of using them in a possible Weekend World programme on the sanctions story. But surprisingly London Weekend failed to appreciate the significance of the documents, and showed little interest in the incriminating evidence that had fallen into their hands. The result was that nearly a full year went by before the Sandford papers were actually used by the media.

Gerry Gable got a very different reaction when he showed Rowland a copy of the Sandford file. Rowland, who became extremely excited by the material, immediately obtained a copy of the papers from Gable, realising that the documents would form a vital element in Lonrho's legal case against the oil companies.

Shortly afterwards the Foreign Office was to learn that the evidence against the British oil companies was considerably more damaging than had been envisaged when the official inquiry had first been established. For in August 1977 Rowland invited the head of the Foreign Office's Rhodesian Economic Department out to lunch. D. I. Mellor, who headed the euphemistically-titled unit responsible for sanctions, was briefly shown the Sandford File – and became very disturbed by what he learned. But Rowland was only willing to hand the papers over on condition that they would not be passed on to the Bingham Inquiry. It will be remembered that Rowland, who was angry about the way the inquiry was operating, had refused to assist the official investigation.

At the end of lunch, Rowland suddenly walked off, leaving the Sandford File on the table. Mellor picked it up, took it back with him, and handed it to Dr Owen. So furious was the Foreign Secretary to discover the extent of the oil companies' involvement in sanctions-busting that he even wondered whether, in the light of such incriminating evidence, the Bingham Inquiry should continue at all. There was a strong argument for simply handing over the whole business to the Director of Public Prosecutions to consider legal proceedings against Shell and BP. Dr Owen was also worried that the Sandford documents might be published in the press, and that this would spark off another major diplomatic row with black Africa.

In the end, however, the Sandford papers made their first public appearance in an unlikely source. Bernard and I had already realised that Jorge Jardim was one of the key people in the sanctions story. Jardim had not yet publicly spoken out against the oil companies, but we knew that from his position as head of the Sonarep refinery at Lourenço Marques he was certain to have a great deal of direct knowledge about Rhodesian supplies.

By this time Jardim had fled from Mozambique, and was based in the Canary Islands. Bernard and I tried to track him down, but he travelled a great deal and proved a difficult man to reach. Finally we discovered through a Portuguese contact that when Jardim visited London he normally stayed at a suite on the ninth floor of the Churchill Hotel. After phoning the hotel every few days, on the evening of Saturday 24 June 1978, I suddenly found myself speaking to Jardim.

At last, I realised, we were near to finding a witness who might be willing to speak openly about this hitherto taboo subject. 'I would be delighted to meet you to discuss how the oil companies bust sanctions', Jardim explained, 'but I think that you ought to know that my whole story will be appearing in tomorrow's *Observer*'. Naturally, our first reaction was one of disappointment. After many months of trying to trace him, another journalist had beaten us to the story by a matter of hours. But when we woke up on 25 June to see Colin Legum's front-page story – headlined 'Oil chiefs bust sanctions' – any resentment we might have felt quickly turned to astonishment.

Until this point public knowledge of oil sanctions-busting was largely confined to allegations suggesting that the South African subsidiaries of the oil companies had been supplying Rhodesia through Freight Services. Suddenly the dimensions of the story had dramatically widened. Extracts from Jardim's documents showed that both the head offices of the oil companies and the British Government themselves had been deeply implicated in the sanctions scandal for over a decade.

The *Observer* article quoted extracts from the book that Jardim was about to publish, *Sanctions Double-Cross: Oil to Rhodesia*. At the time nothing was revealed about the original source of Jardim's documentation, and how it had fallen into his hands. But it is now clear that the papers which Jardim had obtained were none other than the 41 pages of the Sandford File. These had passed through the 'whistle-blower of Britannic House', an intermediary, Gerry Gable, and Rowland, with Jardim finally obtaining the file from Rowland in return for some of his own material. 'We did a little swap arrangement of our own,' Jardim later recalled.[14] It was through this complicated chain that the secret BP documents came to be released in a book published in Lisbon by a Portuguese businessman.

The material in the Sandford File showed conclusively the lengths to which Shell and BP had gone to cover up their trade with Rhodesia. As such, it was political dynamite. It was therefore surprising not only that such sensitive information had been put on paper in London, but also that copies had then been retained in the BP files. Arthur Sandford had been the BP Regional Co-ordinator for Africa from February 1967 until his retirement in September 1975, and had dealt with the tricky Rhodesian question for most of the period since UDI. Along with his counterparts at Shell, Dirk de Bruyne and John Francis, Sandford was therefore one of the main characters in the sanctions story, and the 41 pages of his file clearly revealed the main outlines of the oil scandal.

The first document, a 'strictly confidential' file note dated 1 January 1968, provided an account of a meeting between Shell and BP in London to discuss what action should be taken to satisfy the Government that British-owned oil companies were not busting sanctions. At one point in the discussion the Shell Regional Co-ordinator for Africa, Dirk de Bruyne, declared himself to be 'most reluctant to take any steps which might be construed as an attempt from London to impose limitations upon the trading operations of the [joint Shell/BP] Consolidated Area marketing companies in South Africa and Mozambique'.[15]

The Sandford File then showed that just a few days later Shell and BP had discovered that their jointly-owned company, Shell Mozambique, had been illegally handling supplies for Rhodesia. The papers revealed that a tussle had soon developed between Shell and BP as to whether this embarrassing news should be relayed to the British Government. At a further meeting on 21 February 1968 de Bruyne at one stage 'cut in to say that there must be no mention of the arrangement under which these leakages [to Rhodesia] could take place' (B6.70). Later that day, however Commonwealth Secretary George Thomson was told part of the story by the oil companies and the Government was informed that Shell Mozambique had been busting sanctions. Hints were also dropped that the oil companies had subsequently set up a swap arrangement with Total to supply their Rhodesian subsidiaries.

Later documents in the Sandford File outlined exactly how the swap arrangement with Total operated. A memorandum by Shell executive John Francis dated 4 February 1969 explained: 'Basically there is an exchange arrangement between CFP [Total] on the one hand, and Shell and BP on the other hand' (B/A 267). This was quite obviously a cosmetic device designed to allow Shell and BP to continue to supply their Rhodesian companies.

There was then a five-year gap in the Sandford File before a series of incriminating papers explained the supply procedures which operated in 1974. By this time the swap arrangement with Total had been ended, and Shell Mozambique was once again illegally handling oil for Rhodesia. A 'strictly confidential' BP memorandum on 'Rhodesia and Freight Services', prepared in Cape Town by John Rounce in February 1974, began by giving a concise explanation of the paper-chase:

> BP and Shell continue to market products in Rhodesia as a consolidated venture. Supplies to support the marketing activity are effected from South Africa primarily through 'Freight Services', who act as forwarding agents, buying product from BP and Shell SA and reselling to the Rhodesian Government Procurement Agency Genta for allocation to marketers (B8.39).

The 'Rounce Memorandum' then went on to explain that 'devious supply arrangements have thus been made to visibly disassociate the oil companies from any first hand and identifiable part in supply operations'. This explanation was remarkably similar to the description of supply procedures I had given in *Shell and BP in South Africa*, which had been so categorically denied by oil company spokesmen in London.

The next document in the Sandford File was particularly

incriminating – an internal BP note dated 12 March 1974, part of which read: 'Clearly the arrangements, as they stand, are far too transparent and changes will have to be made' (B8.47). The note added that the head of Shell Rhodesia was shortly due to visit London and that this would provide 'an opportunity of discussing our ideas with him'. Eight days later Ian McCutcheon, a British executive working for BP in South Africa, came up with a plan: 'We will arrange that any correspondence with Genta will not appear in our Freight Services' files or any of our official files. We will also arrange that there is some backdated correspondence agreeing on volume and prices of products moved since 1 January 1974' (B8.48).

In retrospect it is clear that the leaking of the BP documents had two important effects on the breaking of the story. First of all, after the Sandford File had fallen into the hands of the press, the oil companies had little choice but to tell the whole story to the Bingham Inquiry. BP, it is true, have subsequently claimed that their decision to submit all the relevant documents and a full memorandum of facts 'had been taken long before BP learned of the possibility that Lonrho had obtained some of these documents'.[17] Certainly, with the Government holding a controlling shareholding, BP would have been in an extremely difficult position if it had not told the whole truth, and as we have seen, Tom Jackson one of the Government Directors, was also pressing strongly for a full disclosure. Shell, which was not under the same constraints, may initially have been somewhat less inclined to reveal the full facts to Bingham. But after the leaking of the Sandford File, it was obvious that the oil companies could no longer even consider withholding any substantial information.

The second major effect of the leak was that it would now have been much more difficult for the Government to refuse to publish the Bingham Report, since this would have immediately led to cries of a 'cover-up'. It later emerged that Dr Owen, together with many of his Cabinet colleagues, wanted the Bingham Report to be published. There were, nevertheless, a few Ministers who were hesitant. But after so much embarrassing material had leaked out, the Government had little choice but to go ahead and publish. The extensive press publicity about the BP documents also meant that the Bingham Report was finally published rather more quickly than would otherwise have been the case.

It is known that the BP employee who leaked the Sandford File – 'the whistle-blower of Britannic House' – at one point considered releasing considerable further information. This would have included the companion Fraser and Dummett files which contained the papers of Sandford's Managing Directors; a collection of invoices of BP sales to Freight Services which had been kept at Durban; document number 750101, the BP Marketing Operating Plan for 1975; notes taken by R. W. Bentham of BP's Legal Department on the thirteen BP witnesses who ultimately testified to the Bingham Inquiry; and a report of BP's investigation in Mozambique in February 1978. Perhaps the most unusually titled file in BP's Rhodesian collection was one on 'Materials regarded as not relevant'.

*   *   *   *

In the end, however, the 'whistle-blower' struck just once more, this time leaking a copy of the 102-page 'Memorandum Submitted by the British Petroleum Company Limited on 27th September 1977'. This report, which had already been presented by BP to the Bingham Inquiry, was an astonishingly frank admission of how BP's head office in London had discovered that its Southern African subsidiaries had been busting sanctions, and what contacts there had subsequently been with the British Government. Again, however, the document was to remain for many months in the hands of the media before any journalist appreciated its significance.

Ironically, the introduction to the BP Submission included a specific request that Bingham should be particularly vigilant in ensuring that the document remained confidential. Other secret BP documents had already fallen into Lonrho's hands, the Submission pointed out, and these could well be used in a law case against the oil companies. 'This fact highlights the value which BP attach to the assurances of confidentiality contained in the [Bingham] Inquiry's request for information, assurances which are particularly important to BP in view of the existence of these proceedings [by Lonrho] as well as for a wide variety of commercial reasons.'[18]

After the leaking of the BP submission, Britannic House, one employee told us, was like 'the last days of the Nixon White House'. A firm of industrial espionage consultants was brought in to advise; wooden filing cabinets were quickly moved out and replaced with more secure metal cabinets; and executives were to be seen bent over photocopying machines after an instruction had been issued not to allow confidential documents to be copied by secretaries.

Gerry Gable obtained a copy of the BP Submission from his intermediary in March 1978. But again London Weekend Television was strangely unenthusiastic about using the material. Following Jardim's disclosures to the *Observer* on 25 June, however, London Weekend again considered doing a programme based on both the Sandford File and the BP Submission. But by this time Weekend World, the main current affairs programme, had come off the air for the summer break and there was no suitable slot for the BP material.

By this time a rival company, Granada Television's World in Action, had decided to go ahead with its own programme on the sanctions story. Gable therefore took the documents to Granada. World in Action did a two-part special, screened on 31 July and 7 August 1978, which provided the first televised account of how Shell and BP had broken sanctions, and how the Government had condoned these operations. But although Granada had obtained a copy of the BP Submission, little use was made of the astonishing material it contained, and in one hour of television time, just five sentences from the document were actually quoted.

Unfortunately the World in Action programmes, like the *Observer* article a few weeks earlier, had sparked off little follow-up in the press. In fact the television programme seemed to have almost aroused more interest in Lusaka than at home, owing to the Zambian law case against the oil companies. This led to an amusing incident. After watching the programme, the Zambian High Commissioner in London told Granada that she thought President Kaunda would like to see the film. Feeling rather flattered, World in Action dug out a cassette recording and sent it off to Lusaka. Kaunda found a video machine, assembled his entire Cabinet in front of the screen, and switched on. To their surprise the Zambians then found themselves watching a riveting programme about rock-climbing in northern Britain.

Although Bernard and I had worked as consultants on the World in Action programme, we had not been shown copies of the BP Submission at the time. A few weeks later, however, Granada Television generously provided us with a copy of both the BP Submission and the Sandford File. On reading the BP Submission we were astonished at the extent of the knowledge of the London head office of the company. What a contrast, we thought, with the string of public denials that BP had issued to the press: here was an oil company openly admitting its own involvement in sanctions-busting – quite a different thing from partially-substantiated allegations from outside observers.

Our first move was to prepare a detailed commentary on the BP Submission. We then took the Submission and our commentary to Peter Kellner at the *Sunday Times*, who was one of the few journalists we had found who had really appreciated the significance of the sanctions saga. 'At first I thought that the BP document would do little more than add detail to what we already knew,' Kellner later recalled. 'But as the week wore on I realised that we had an unparalleled tale of corporate and Government deceit. We had quite remarkable evidence of the complicity of Ministers and civil servants – a story which exposed how Britain's closed system of government could conceal a scandal of international proportions.'[19]

*　　*　　*　　*

On 27 August 1978 the *Sunday Times* ran a front-page lead under the headline 'BP confesses it broke sanctions – and covered up'. In it we told how Britain's largest oil company, with its Government shareholding, confessed that for most of the past twelve years its subsidiaries had been knowingly supplying oil products to the illegal regime in Rhodesia. The BP Submission admitted the following:

*The head offices of BP and Shell have for ten years known and condoned the fact that their South African subsidiaries were providing half of Rhodesia's oil.

*The two oil giants learned in 1968 that their London-registered, jointly-owned Mozambique subsidiary, Shell Mozambique, had also been involved in the Rhodesia supply chain. This, as they realised, was in direct contravention of UK sanctions law.

*BP and Shell, with the informal approval of the Wilson Government, then arranged for a French company to send oil to Rhodesia on their behalf – in precisely the same quantities as before and on terms negotiated directly between themselves and the Rhodesian state purchasing agency, Genta.

*In 1974 BP learned in explicit detail how their South African subsidiary had resumed direct supply of oil to Rhodesia, with Shell Mozambique, still UK registered, again illegally involved. BP did nothing to stop this and did not inform the government.

*BP's central planning department in London was sufficiently well informed of the company's Rhodesian business to take it into account while planning the expansion of the group's South African refinery in 1975.

*After the setting up of an official inquiry, by Thomas Bingham, QC, last year, BP finally did what it had previously claimed was legally impossible; it forced BP South Africa to end all sales to the South African intermediary, Freight Services, through which it had been funnelling the Rhodesian supplies.

It was *The Sunday Times* revelations which really sparked off public concern over what quickly became known as the Oilgate scandal.

NOTES
1 *Daily Mirror*, 9 April 1977.
2 *Financial Times*, 24 May 1977.
3 *Financial Times*, 24 May 1977.
4 Granada Television *World in Action*, 31 July 1978.
5 Submission to British Government Inquiry on Allegations of

Sanctions Busting by Shell and British Petroleum.

6 *Guardian*, 8 June 1977.
7 Press conference, 12 August 1977.
8 *Guardian*, 17 May 1977.
9 Press conference, 10 June 1977.
10 Letter from Rowland to Dr David Owen, 26 May 1977, p. 1.
11 Letter from Rowland to Dr David Owen, 26 May 1977, p. 1.
12 Press conference, 15 June 1977.
13 *Guardian*, 18 June 1977.
14 Interview, 10 February 1979.
15 File note prepared by R. Bishop (BP) (unpublished).
16 Subsequently some of the Sandford File documents were quoted in the Bingham Report.
17 *New Statesman*, 15 September 1978.
18 BP Submission to the Bingham Inquiry, 27 September 1977, p. 6.
19 Interview, 18 April 1979.

5 Bingham And Beyond

Bingham handed over his report to Dr Owen on 22 August 1978. His investigation had been completed in sixteen months, a remarkable achievement considering the scope of the inquiry. Earlier fears that the small size of his team would prove a disadvantage had proved unfounded. Unfortunately, however, the completion of the Bingham Report was given no more than a few lines in the British press. Interest in the sanctions story seemed to have virtually fizzled out.

It was just five days later that the *Sunday Times* published its dramatic report on the BP Submission. The sanctions saga then hit the front pages of all the British newspapers: 'A scandal in many ways more grave than Watergate,' the *Daily Express* cried.[1] The *Daily Mail* called it 'the major political scandal of post-war Britain'.[2] *The Times* ran an editorial pointing out that 'the issues now extend far beyond the Rhodesian sanctions themselves . . . they involve the integrity of Government'.[3] Suddenly all the press was clamouring for the speedy publication of the Bingham Report.

The oil companies now felt that they would face an even more embarrassing situation if the Bingham Report was not published and the speculation continued. On 31 August 1978 *The Times* carried a letter from BP Chairman Sir David Steel announcing that he would welcome the release of the report, 'it being my confident belief that it will contain a fair and objective account of events'. Meanwhile the impact of the *Sunday Times*' story had hit the oil companies where it hurt most – financially. Within four days 20p had been knocked off the prices of Shell shares, with 28p off those of BP. Altogether this meant that the two oil companies had, albeit temporarily, lost a total of £220 million off their shares as a result of the sanctions disclosures.

The embarrassment faced by the oil companies was illustrated by Sir David Steel's decision on 7 September to issue a special statement to all his staff. 'I am very sorry indeed that you and your families and friends should have been put to all this worry and concern over the allegations and insinuations against BP,' he wrote. 'I still ask people to be patient and refrain from judgement until the full facts are known and fairly considered.' There had been few occasions in recent years in which major multinational companies had come under such sustained attack in Britain.

Shell and BP did react slightly differently to their past involvement in the sanctions story. Shell appeared particularly complacent. Michael Pocock, Chairman of Shell, told shareholders at the 1978 Annual General Meeting – long after he knew how deeply Shell South Africa had been involved in supplying Rhodesia – that 'We have no reason to feel ashamed of the record and action of our subsidiary company in South Africa . . . I feel proud of them'.[4]

BP, while justifying its actions after the publication of the Bingham Report, did admit that 'it would be surprising, in retrospect, if there had not been over the thirteen years, errors of judgement and procedures'.[5] Because of the Government's shareholding in BP, the company was in a particularly sensitive position. After the Bingham revelations Energy Secretary Anthony Benn pushed for tighter government control of the company. Trade unionists within BP also called for a worker-director who, they claimed, would be more responsive to the public interest.

The Government has still not resolved the question of how close a control it should exercise over BP. At present the BP shareholding is actually held by the Treasury, which is not particularly concerned with oil matters, and there has therefore been some pressure for the shares to be transferred to the Department of Energy. A decision has also not been reached on the degree of control which the two Government-appointed Directors should apply. Indeed, since Lord Greenhill's resignation from the BP Board in November 1978 (which was apparently planned before the sanctions row) the Government has not yet nominated a successor.

*   *   *   *

But it was not only the oil companies that were blackened by the disclosures. The Sandford File and the BP Submission had revealed that the existence of the Total swap arrangement had been known to the authorities a decade earlier. A Labour Government had therefore condoned the activities of Shell and BP in supplying Rhodesia. In September 1978 normally confident political figures suddenly quivered at the very mention of Rhodesian oil. Chancellor of the Exchequer Denis Healey lost his temper during a visit to Canada when he was questioned about the Bingham revelations and shouted at reporters, 'I am not prepared to be interrogated as if I were a criminal in the dock.'[6] A few days later Wilson actually broke off a television interview in anger after being questioned about the sanctions scandal. 'I'm very sorry. I'm very sorry. Will you stop the camera,' he said, pushing his hand in front of the camera. 'I'm very sorry. I really cannot be treated like this.'[7]

Thomson, as the Minister who had led the Government's discussions with the oil companies in 1968–69, and had therefore learned about the Total swap arrangements, became the main target of public attack. On 7 September 1978 Thomson took the unusual step of issuing a detailed personal statement explaining the background to the sanctions story. 'Since it is dangerous to rely on memory for events of ten years ago,' he began, 'I have exercised my rights as a former Cabinet Minister to consult the appropriate papers.' He then went on to claim that 'they confirm that I conveyed in writing to the Prime Minister and other Ministers most directly concerned a full account of all that passed at my meetings on behalf of the Government with the oil companies'.

Wilson replied the following day with a rather unclear statement of his own, suggesting that he had never been informed about the Total swap arrangement. The ambiguity of his comments was reflected in the very different headlines in the *Guardian* ('Wilson says he knew of sanctions-busting') and *The Times* ('Sir Harold denies any knowledge of oil details').[8] Soon afterwards William Hamilton MP called Wilson the 'Houdini of Huyton'.[9]

Wilson also managed to implicate James Callaghan by claiming that he was among Ministers who 'have expressed their inability to remember hearing the allegations' about the Total swap. The following day *The Times* added to the confusion by reporting that 'sources close to Mr Callaghan yesterday were "mystified" by his inclusion as a minister having either talked to Sir Harold or expressed an opinion in any other way'.[10] It now seemed quite impossible to sort out the truth from all the jumbled statements in the 'who knew what' debate.

To the public the sanctions affair appeared to confirm their view of politicians as cynical, untrustworthy figures. Many people probably shared the views of one angry reader who wrote to the *Sunday Times* to give his reaction to the oil fiasco: 'At the next election make it a rule that no person who has ever before been a Member of Parliament should be eligible to stand as a candidate again.'[11] Such was the increase in public cynicism over the Labour Government's policies that some political commentators suggested that the sanctions scandal may well have been one of the reasons behind Callaghan's decision not to call an election in October 1978.

*　　*　　*　　*

The Foreign Office holds a daily press briefing which is usually devoted to little more than listing the formal engagements of Foreign Office Ministers. But at 12.30 pm on 13 September 1978 the normally uneventful meeting was dominated by a two-sentence statement:

> The Government have decided that the report of Mr T. H. Bingham QC and Mr S. M. Gray FC on the supply of petroleum and products to Rhodesia should be published in full at the earliest possible date as soon as consents to disclosure have been obtained in accordance with the requirements of the Southern Rhodesia (United Nations Sanctions) (No. 2) Order 1968. The Government have also concluded in the light of material in certain of the Report's findings that it should be referred to the Director of Public Prosecutions to consider whether further inquiries should be undertaken with a view to possible criminal proceedings for breaches of the Sanctions Order.

It is clear that the *Sunday Times* disclosures had forced the Foreign Secretary to 'study' the Bingham Report as quickly as he did, to submit it immediately to the DPP, and to opt for immediate publication.

Her Majesty's Stationery Office was on strike at the time, and the Cabinet itself therefore had to take the decision to have the report reproduced for distribution. The first edition was poorly printed, from a typed copy, and was unbound. At precisely 3.30 pm on Tuesday 19 September the Bingham Report was officially released. The timing had been carefully arranged so that 'price-sensitive' information would not be released until the Stock Exchange was closed for the day. Privileged reporters, however, were admitted to a special room at the Foreign Office where they were allowed to study a copy of the report two hours before official publication, under the stern gaze of an attendant who would not let them leave before 3.30.

The Bingham Report is a difficult document to digest, written in a dense, legalistic terminology, and displaying great subtlety of language and judgement. But there is no doubt that the report provides a remarkably comprehensive account of the involvement of Shell and BP in Rhodesia. The body of the report is over 100,000 words and is accompanied by nearly a hundred pages of supporting documentation. But with so little time in which to prepare their stories, none of the reporters were able to do much more than reproduce the conclusions of the inquiry. Even today few journalists have actually read the Bingham Report from cover to cover.

One part of the Bingham Report has never been published – Annex III on 'Evidence of Criminal Offences'. It is a short document, just several pages long, which names the oil company executives who may have committed offences against the Sanctions Order. Together with the main report, this annex had been handed to the Director of Public Prosecutions on 4 September 1978. Publication of Annex III, however, was withheld at the request of the DPP 'while matters covered by the report are under consideration by him'. Within a few weeks the DPP had received preliminary advice from Counsel and had arranged for Chief Superintendent Jim Smith, a senior officer from Scotland Yard's Fraud Squad, to investigate whether criminal proceedings should be initiated. On 27 April 1979, after Scotland Yard's preliminary report had been completed, both Shell and BP were served with notices ordering them to deliver all documents concerned with Rhodesian oil supplies to Scotland Yard.

*   *   *   *

Publication of the Bingham Report also sparked off yet another row between Lonrho and the Government. On 18 September 1978, the day before the inquiry's findings were released, Rowland wrote a vitriolic letter to the Foreign Secretary making a very serious allegation. He claimed that Government Ministers had been party to a damaging attempt to harass Lonrho as a form of retaliation against Lonrho's decision to sue the oil companies involved in sanctions-busting.

Rowland's letter recounted some of the events of the previous year. On 6 July 1977 he had talked with the Foreign Secretary for more than an hour about Rhodesia. The two men began by discussing the allegations contained in the Department of Trade report that Lonrho itself had been involved in sanctions-busting. Rowland apparently took the stand that he was unwilling even to talk about other Rhodesian matters with the Foreign Secretary while the possibility of legal action was still hanging over his company. He claims that he then received a personal assurance from Dr Owen that a statement clearing Lonrho of all charges 'would be made during the recess or immediately Parliament reassembled'.[12]

Dr Owen, who has disputed Rowland's account of the meeting, has argued that he was merely trying to encourage Lonrho to cooperate with and give evidence to the Bingham Inquiry. But the Lonrho chief has maintained that the Foreign Secretary also had a further purpose in meeting him – to put pressure on him to drop his company's law case against Shell and BP. It was 'your polite but strong contention', Rowland later told Dr Owen, 'that the case ought to be withdrawn.'[13]

Lonrho, however, continued to press forward with legal action. Shell and BP wanted the dispute to go to arbitration, rather than being heard in open court, and the oil companies took this issue to the High Court in London. On 31 January 1978 Mr Justice Brightman ruled in favour of Shell and BP.

Lonrho's allegations of breach of contract, he determined, should be settled by arbitration. The arbitration proceedings have begun, but a judgement has not yet been reached.

At the time of the publication of the Bingham Report Lonrho had still not yet received the statement which Rowland had apparently been led to expect by Dr Owen, giving the company a clean legal bill of health. Indeed, signs to the contrary had already emerged. On 4 May 1978, Lonrho executives were surprised to receive a visit from the City of London Fraud Squad armed with a warrant, signed by the Foreign Secretary, demanding the production of various documents relating to Lonrho's dealings with Rhodesia. News of the visit was leaked to the press one week after, at the very worst moment for Lonrho, when the company had just made a bid to acquire Suits (Scottish and Universal Investments). On the following day, after press reports had appeared about the Fraud Squad's visit, Prices Secretary Roy Hattersley announced that the Lonrho bid for Suits would be referred to the Monopolies Commission. It was Rowland's belief that the British Government was now making a concerted attempt to harass his company because of his activities in uncovering the sanctions scandal.

But worse was still to come for Lonrho. Just under a month later, on 2 June 1978, the Tanzanian Government suddenly announced that it was nationalising all of Lonrho's assets in the country. Two reasons were given. Rowland apparently had not hesitated 'to meddle in the politics of Southern Rhodesia', the Tanzanians claimed, and the UK Department of Trade report had shown that Lonrho 'while posing as a friend of free Africa, has over a number of years engaged in profit-making activities in Rhodesia' in a way inconsistent with UN sanctions.[14] Rowland later suggested that Tanzania's retaliation had been taken with the implicit approval of the British Government. 'The Foreign Office,' he told Dr Owen, 'have made it clear to the Tanzanian Government that to cut us off with a Tanzanian shilling was unobjectionable.'[15]

At the time of the publication of the Bingham Report relations between Lonrho and the British Government had reached their lowest ebb. Rowland decided to publicise the dispute by releasing copies of his correspondence with Dr Owen, which were published in *The Times* on 3 October

1978. Two days later the newspaper carried an editorial arguing that the Foreign Secretary should be 'required to answer the charges more comprehensively than he has chosen to do so far'.[16] The row seemed set to escalate still further. Lonrho seemed about to release further incriminating material on the sanctions scandal. On 22 October *The Times* placed an advertisement for itself under the heading 'Who's better informed about Rhodesian oil than Ian Smith?' Rowland, was their answer. 'Further issues of the debate will be raised in *The Times* over the next few intriguing days,' the newspaper promised. But nothing actually appeared.

Lonrho's row with the Government ended suddenly. On 3 November 1978 the company issued a brief announcement that the Director of Public Prosecutions had decided not to press any charges arising out of the Department of Trade inquiry. There was some speculation that the timing of the DPP's decision was not unconnected with the fact that Parliament was due to debate the Bingham Report just a few days later. The Government was presumably only too aware that if Lonrho itself had not been cleared, then Rowland might have been tempted to make further incriminating allegations about the Foreign Office's involvement in the sanctions affair.

Some evidence to confirm this view emerged when the House of Commons debate later took place. Edward du Cann MP, who is also a Director of Lonrho, told Parliament that he hoped the Foreign Secretary would 'make information available about the situation when Mr Rowland was talking to one of his officials in the Foreign Office, Mr Mellor'.[17] Du Cann suggested that this had been a meeting of some significance, probably referring to the occasion when Rowland gave the Government a copy of the Sandford File.

*   *   *   *

The Bingham revelations also had an important impact overseas. 'The affair,' the *Sunday Times* predicted 'threatens to do more damage to our battered international reputation than anything since Suez.'[18] But it was in Zambia, which had suffered so much from UDI, that the revelations caused the biggest stir. 'I am so angry that I cannot describe how I feel,' President Kaunda fumed at reporters. Successive British Prime Ministers, he claimed, had 'lied' and 'cheated' over what he called the Oilgate scandal. It was worse than Watergate, he added, because it had already cost the lives of thousands of Rhodesians.[19]

Two days after the publication of the Bingham Report a terse announcement was issued by Number Ten. James Callaghan would be flying to the northern Nigerian town of Kano for discussions with Kaunda. 'Action Man Jim zooms off to win back trust,' 'Jet-away Jim goes on "cool it" mission', screamed the popular press.[20] *The Times* carried a front-page lead report which soberly announced: 'Callaghan move to save British assets from African retaliation.'[21] Fears were expressed in financial circles that Kaunda was seeking to persuade the Nigerian Government to sequester the assets of Shell and BP in Nigeria. Other reports suggested that Zambia might be planning a full-scale nationalisation of British investments there.

The fact that the British press corps was not given the usual facilities to fly in the Prime Minister's aircraft meant that few reporters were able to make independent arrangements to reach Kano in time to cover what promised to be a dramatic confrontation. This merely added fuel to suggestions that Callaghan did not want witnesses at what could be a humiliating climb-down in the wake of the Bingham storm. In the end, however, the Kano meeting ended in accord. Britain promised to supply Zambia with a large consignment of arms. This, many observers remarked, was 'reparation' towards the enormous damage caused by the sanctions débâcle.

The Bingham revelations also caused considerable concern in the countries where the other international oil companies were based. The main terms of reference of the British investigation were confined to an examination of the UK oil companies. Bingham, in any case, lacked the authority to compel foreign oil companies to give evidence to his inquiry. Nevertheless, incidental references to Mobil, Caltex, and Total arose throughout the investigation. All the evidence in the Bingham Report, as well as other published material, has suggested that all five international oil com-

98

panies had used similar methods to bust sanctions. In particular, they all appeared to have made use of Freight Services as an intermediary to supply Rhodesia.

The strong anti-apartheid tradition in Holland made the sanctions scandal the cause of widespread anger there. Shell's subsidiaries in Southern Africa come under the authority of the London office. But Royal Dutch Petroleum, registered in The Hague, has a sixty per cent shareholding in the Shell group, and there was also considerable interchange of personnel between the Dutch and the British Shell companies. The close interlocking ties between the Dutch and British parts of the Shell group meant that The Hague was still deeply involved in supporting Shell's operations in Southern Africa. It was therefore widely believed in Holland that Royal Dutch Petroleum was partially responsible for the sanctions-busting activities of Shell's subsidiaries in Southern Africa.

After the publication of the Bingham Report there was strong pressure for a Dutch investigation into sanction-busting. In December 1978 the Dutch Parliament finally set up its own inquiry. The investigation is being mounted by a special seven-man subcommittee of the Foreign Affairs and Commercial Policy Parliamentary Standing Committee. The inquiry, which is chaired by Christian Democrat MP J. N. Scholten, is due to present its interim report in June 1979. The investigation's terms of reference are to examine the roles played by both the Netherlands Government and by Dutch oil companies and citizens in the supply of oil to Rhodesia.

The three Dutch citizens who have served as Shell's Regional Co-ordinator for Africa will presumably be important witnesses – Dirk de Bruyne (1965–68), J. de Liefde (1972–77), and A. de Geus (from 1978). De Bruyne, who is now President of Royal Dutch Petroleum, was also the Shell Managing Director responsible for Africa between 1972 and 1976, and he is obviously particularly well placed to assist the investigation.

The Bingham revelations also caused a stir in America. In mid-September 1978, just a week before the publication of the sanctions report, US Ambassador to the UN Andrew Young sent a confidential cable to the State Department in Washington urging that a detailed investigation should be initiated into the activities of <u>Mobil</u> and Caltex, the 1976 inquiry into *The Oil Conspiracy* allegations by the Treasury having proved quite inconclusive. If the US Government did not initiate a further investigation, Young argued, then the Anglo-American attempt to work for a Rhodesian settlement could well founder on African suspicions.

In the State Department, however, the situation was viewed rather differently. On 8 November 1978 Assistant Secretary of State Richard Moose told *Africa News* that the Bingham Report had shed 'no light whatsoever on the activities of American oil companies'.[22] Moose added that the Government had 'done everything that we can do within the scope of jurisdiction of the US Government to ensure that American-controlled companies are not involved in this trade'.

Soon after Moose's statement, however, the Assistant Secretary of State was embarrassed to learn that just a few days earlier the US Treasury had in fact decided to reopen its previous investigation. *Africa News* later reported that 'more than one disgruntled official has suggested that the Treasury may have acted to prevent a more independent inquiry. Public pressure, said one, is all that will prevent a new investigation from being less superficial than the last one'.[23]

Meanwhile <u>Mobil</u> claimed that no evidence had emerged to show that its South African subsidiary had been involved in supplying Rhodesia. In September 1978, just before the Bingham Report was released, Vice-President George Birrell told Bernard Rivers, 'We have no basis for believing that Mobil South Africa has ever knowingly used Freight Services as a conduit for supplying petroleum products to Rhodesia'.[24] Birrell then added that he still stood by his testimony to the US Senate given two years earlier.

Even after the publication of the Bingham Report Birrell said that he was still convinced that Mobil South Africa had never bust sanctions. His reasons, he said, were the assurances he received from Durban. 'If we have to choose between what [former Managing Director] Beck tells us and what somebody outside the company is saying about Mobil South Africa, I will choose to believe Bill Beck. In our company, I don't think people lie to each other.'[25]

Nevertheless, once the full facts are revealed, it is likely

100

to be proved that <u>Mobil</u> – like Shell and BP – was indeed supplying Rhodesia through Freight Services. Birrell may then regret his proclamations of Mobil's innocence to the US Senate.

Caltex had taken a similar stand to Mobil over Rhodesian trade. In January 1968 Texaco, co-owner of Caltex, revealed that an investigative team had been sent out to Cape Town to determine whether the company's local subsidiary had been involved in sanctions-busting. Texaco claimed that it had found that 'Caltex South Africa has not since the imposition of sanctions sold oil products to any entity with the intention that these shall find their way to Rhodesia'.[26] Again after the publication of the Bingham Report, Caltex continued to deny vehemently any involvement in sanctions-busting. 'I don't think we are interested in saying more than this at the moment,' a Caltex spokesman added.[27]

So far little attention has been focussed on the role of Esso South Africa, a subsidiary of Exxon, which has a small operation in Southern Africa. But there is evidence to suggest that Esso was probably involved in the sale of lubricants to Rhodesia through intermediaries. This incriminating evidence emerged during a law case in 1971, in which Esso South Africa took a firm known as Virginia Oil and Chemical Company to court over a payments matter. Virginia then claimed that Esso had used the company as an intermediary to supply Rhodesia. The American head office of Exxon, however, later denied this allegation: 'We were certainly not using Virginia Oil and Chemical to make sales to Rhodesia.'[28]

Information from a confidential BP document has also suggested that Esso South Africa attempted to take over the *entire* Rhodesian oil market in 1970 at discounted prices (B8.39). Exxon's US head office replied: 'We have no knowledge of such an offer ever having been made . . . We just feel it has absolutely no basis in fact.'[29] The Exxon spokesman concluded that 'we have not been supplying oil to Rhodesia. We don't need that kind of trouble'.

Since US sanctions legislation also applies to American citizens resident overseas, the new Treasury investigations will focus particularly on US Directors who sit on the boards of the oil companies' Southern African subsidiaries. Since UDI a number of American Directors have sat on the Board of Mobil South Africa (including Fanueil Adams Jr, John Calvert, Everett Checket, and Charles Scott). All of these, however, have been based in the United States, and have claimed to have actually played little part in the management of the South African company. A US Director was also a member of the Board of Mobil Rhodesia until 1969.

The Mobil Director with the closest contacts with South Africa has been Charles Solomon. Although born in South Africa, he became an American citizen in 1963, and has also known William Beck (Chairman of both Mobil South Africa and Mobil Rhodesia until 1978) for over twenty years. Since Solomon was appointed to the Board of Mobil South Africa in 1972, he has visited Durban every year at the company's expense. But the earlier US Treasury investigation reported that Solomon 'insisted that on his annual trips, only one to two weeks are devoted to company matters, the remaining time being spent on vacation'.[30]

The fact that the 1976 Treasury inquiry did not investigate Caltex or Exxon means that much less is known about the involvement of American Directors of these two companies in their South African operations. But at least one American Director of Caltex actually visited Rhodesia recently. Bingham reported that one witness claimed that a US Executive Director of Caltex 'did blow in (to Rhodesia) in a sports shirt and jazzy trousers' (B12.30). In addition the Director of Caltex South Africa responsible for distribution – who is particularly well placed to know about supplies to Rhodesia – is Philip Conlan, who is believed to be a British citizen, and therefore subject to the UK Sanctions Order. Conlan would presumably be a valuable witness for the Treasury investigators.

Bingham's report shows the extent of the documentary material and personal testimony which he managed to obtain in Britain on the roles of Shell and BP in supplying Rhodesia. There is therefore no reason to believe that sufficient evidence does not exist in the United States to prove that the American oil companies were also involved in sanctions-busting. In spite of this the half-hearted attempt by the Treasury investigators in 1976 suggests that it is still possible that the US Government may produce an equally inconclusive report. Whether the full facts about the involvement of the American oil companies in sanctions-busting ever emerge depends ultimately on the question of political will.

Both the French Government and Total have continued to proclaim their innocence, even in the light of the incriminating evidence that emerged during the Bingham Inquiry. On 16 November 1977 the French Representative to the UN issued a formal statement claiming that its investigations had shown that 'Total, not being involved in any way in supplying Rhodesia, is not violating the sanctions in force against that country'.[31] Nine days later Total's Director of Information wrote that 'on information available in Paris, Total South Africa was not in 1968, and has not been since then, a clandestine supplier of Rhodesia'.[32]

This is a strange conclusion in view of evidence revealed in the Bingham Report that Shell and BP had used Total South Africa to handle their Rhodesian supplies between 1968 and 1971 under a swap arrangement. In addition, Total also provided supplies for its own marketing company in Rhodesia. Yet even after the publication of the Bingham Report the French Government again formally told the United Nations that its investigations into Total had shown that 'there were no grounds to believe that the company had been supplying oil to Southern Rhodesia'.[33]

The strong denials from Total's head office in Paris are perhaps surprising, since a number of French executives have detailed knowledge of the company's Southern African subsidiaries. Daniel Banmeyer, the present Chairman of both Total South Africa and Total Rhodesia, and at least five of the Directors of Total South Africa since UDI (Etienne Dalemont, Louis Deny, Jean Flamand, A. Parquet, and R. G. Pomeroy) are French citizens. Louis Deny, who is now Director General of the Compagnie Française des Pétroles (Total) in Paris, was also Managing Director of Total South Africa from UDI until 1967. He was then replaced by Jean Flamand, who held the post until 1972. There are therefore a number of key Total executives now based in France who have direct knowledge of the activities of the Southern African subsidiaries of Total. This makes it difficult to understand how Total can claim that it has no information to suggest that its South African subsidiary has ever supplied Rhodesia.

The fact that the French Government has forty per cent

voting rights in Total might have been expected to have made the sanctions issue particularly sensitive. But French companies have continued to maintain very close links with the white regimes of Southern Africa, and right from the early days of UDI the French Government has never been enthusiastic in its application of sanctions. There has therefore been no attempt by the French authorities to investigate evidence of sanctions-busting by Total in the light of the Bingham revelations.

* * * *

It was in Britain, of course, that the Bingham Report had its most important political repercussions. It was perhaps fortunate for the Labour Government that the *Sunday Times*' revelations and the actual publication of the Bingham Report took place during the Parliamentary recess. Had MPs been at Westminster when the Bingham Report was released, there would no doubt have been angry calls for an immediate investigation into the Government's handling of the affair. As it was, one MP, William Molloy, demanded the recall of Parliament to discuss the oil embargo after the publication of the *Sunday Times*' article.

It was not until the annual November debate on the Sanctions Order that MPs had the chance to discuss the Bingham Report in the House of Commons. The debate, however, was combined with a wider discussion of the politics of the current Rhodesian situation and the need to renew the Sanctions Order. *The Times* reported that the Government had reached 'an unusual arrangement with the Opposition to wrap up . . . both the present situation in Rhodesia and the past years of oil sanctions-breaking . . . While it might strike the casual onlooker as perfectly logical to take all the Rhodesia matter together, the arrangement smacks of a political fix to some observers.'[34]

On 5 November 1978, shortly before the Parliamentary debate, Social Audit, a pressure group 'concerned with improving government and corporate responsiveness', published *A Review of the Bingham Report*. The analysis, written by Andrew Phillips, extracted some of the most important evidence contained in the Bingham Report on the interaction of the oil companies and the British Government.

104

Social Audit explained that it had decided to publish Phillips' review 'not to be sanctimonious about past mistakes or lies or dishonesty – involving Ministers, civil servants, and Shell and BP – but to inform the public and their elected representatives how they were deceived, repeatedly and continuously, over a period of more than twelve years.'[35]

Wilson was the most eagerly awaited contributor to the Parliamentary debate. Over the previous few weeks it had become increasingly clear that his government had played a devious role over Rhodesian sanctions in condoning the Total swap. Indeed, the *Rhodesia Herald* had jokingly offered the former Prime Minister a home in Salisbury: 'Perhaps it would be stretching things a bit to suggest that Sir Harold was, in fact, our man in Whitehall. But if things do get hot for him at Westminster, should we not let bygones be bygones and offer him political asylum?'[68]

When Wilson spoke before the House of Commons on 7 November 1978 he gave a long and complicated explanation of why he and his advisors had never learned about the existence of the Total swap arrangement in 1968–69.

At one point Roderick MacFarquhar, MP, cut in to point out that 'there must somewhere have been a grotesque error of judgement,' and that it was impossible to see, 'in his catalogue of all the people who could not have been guilty of that grotesque error of judgment, where it lay'.[37] Wilson then placed the blame firmly on the shoulders of Thomson: 'I do not believe that when my noble Friend heard what was said [by the oil companies in 1968–69] he realised the implications.'[38] Thomson, who was sitting in the gallery of the House of Commons, glared down at his former Prime Minister. Wilson then lamely concluded that 'Oil is a viscous fluid, and nowhere is it more viscous than in South Africa.'[39] Again, his contributions had raised more questions than it answered.

Two days later Thomson had his chance to reply in the House of Lords. He once more tried to dispose of claims that he 'was conducting some private Rhodesian policy of my own with the oil companies'.[40] The records of his two meetings with Shell and BP in 1968–69, he pointed out, had been sent to Number Ten. He then justified his decision to accept the swap arrangement with Total. 'The Government's record,' he claimed, 'was a creditable and honourable one.'[41]

There continued to be considerable pressure for a further investigation into the sanctions story. Bingham, it was widely acknowledged, had provided a comprehensive account of the activities of the oil companies, but he had regarded it as beyond his brief to make any detailed investigation into the role played by the British Government. Thomson, after all, had been the only politician to be interviewed by the Bingham Inquiry. The importance of the issues raised by the Bingham revelations went much further than Rhodesia, as many MPs realised, and there was therefore the need for a further investigation to examine the Government's handling of the sanctions scandal.

Differences quickly arose within the Cabinet on the need for a further inquiry. Some Ministers would have liked to have seen the establishment of a Parliamentary investigation. Others realised that a further inquiry would inevitably reveal some embarrassing information which would hardly help the Labour Party in a run-up to an election. They argued that the past should be left alone, because the Bingham Report had uncovered the main outlines of the story, and all efforts should be concentrated on the present – on reaching a settlement to the Rhodesian situation. This argument, however, missed the fundamental point – that the sanctions story raised issues concerning the way the British system of government worked.

On 15 December 1978 Callaghan told the House of Commons that 'the exceptional nature of the events covered by the Bingham Report and the importance of the questions raised by hon. Members calls for a further inquiry'.[42] The Prime Minister added that 'if we do not have an inquiry, those who wish to believe the worst of us will go on alleging that there is some kind of Watergate here that has to be covered up'.[43] Callaghan then outlined the terms of reference of the proposed Parliamentary inquiry:

> To consider, following the Report of the Bingham Inquiry, the part played by those concerned in the development and application of the policy of oil sanctions against Rhodesia with a view to determining whether Parliament or Ministers were misled, intentionally or otherwise, and to report.[44]

It was a very unusual situation. Callaghan himself admitted that the last time a Special Commission of both Houses of Parliament had been set up was after the Dardanelles Campaign in World War I. The new inquiry was unprecedented in having access to Cabinet papers, which are not normally made available for thirty years. Agreement had been obtained from the previous two Prime Ministers (Wilson and Heath) for these papers to be available to the Chairman of the investigation, who alone would have the right to sift the documents, and decide which should be made available to the inquiry. The Special Commission on Oil Sanctions, as it was to be titled, would be an eight-member team consisting of members of both Houses of Parliament, chaired by a Lord of Appeal. The inquiry would sit in private, but its findings would be published.

Some MPs still feared a cover-up. Hearings would be in private, and Cabinet papers would initially be only seen by the Lord of Appeal. But the arrangements proposed were a reasonable compromise to deal with a complicated and unprecedented situation. On 1 February 1979, the House of Commons debated a motion to establish a Special Commission on Oil Sanctions. Most of those who voted in favour were Labour MPs, although they were joined by a few Conservatives, and Liberal MPs also backed the proposal with great vigour. Liberal leader David Steel told Parliament: 'These issues cannot just be swept aside as being of purely academic interest to future historians.'[45] By a free vote the motion was approved by a substantial majority, 146 to 67. One week later, however, a similar motion was rejected by the House of Lords by 102 to 58. It was unusual for the House of Lords to overthrow a decision of the Commons, and once again it raised the thorny question of Parliamentary reform. But the immediate effect was to kill the proposal for a Special Commission of Inquiry.

On 12 February 1979 Leader of the Commons Michael Foot announced that he 'deeply regretted' the decision of the House of Lords, and promised that 'the Government will come forward with proposals for dealing with the situation'.[46] But the sanctions scandal became buried by more immediate problems – first the winter wave of strikes in Britain, then the devolution referendum, and finally the May election.

Bingham's investigations had unearthed the story of the

oil companies' involvement in sanctions-busting. But the investigation had raised more questions than it had answered. Strong evidence suggested that both Ministers and Parliament had been misled over the most important foreign policy issue which Britain has faced since Suez. In its implications the Bingham Report therefore went much further than the Rhodesia problem – important though that was. It raised serious questions about the whole system of government in Britain. Until a further Parliamentary inquiry is actually set up there will inevitably be lingering fears of a further cover-up. What is it that is being kept from us now?

NOTES

1 *Daily Express*, 5 September 1978.
2 *Daily Mail*, 20 September 1978.
3 *The Times*, 7 September 1978.
4 18 May 1978.
5 BP statement, 20 September 1978.
6 *Daily Telegraph*, 22 September 1978.
7 *The Times*, 29 September 1978.
8 *Guardian* and *The Times*, 9 September 1978.
9 Hansard, Commons, 7 November 1978, col. 842.
10 *The Times*, 9 September 1978.
11 *Sunday Times*, 20 September 1978.
12 Rowland's letter to Dr Owen, 18 September 1978 (reprinted in *The Times*, 3 October 1978).
13 Rowland's letter to Dr Owen, 18 September 1978 (reprinted in *The Times*, 3 October 1978).
14 Tanzanian statement, 2 June 1978.
15 Rowland's letter to Dr Owen, 18 September 1978 (reprinted in *The Times*, 3 October 1978).
16 *The Times*, 5 October 1978.
17 Hansard, Commons, 7 November 1978, col. 759.
18 *Sunday Times*, 3 September 1978.
19 *Daily Telegraph*, 19 September 1978.
20 *Daily Mail* and *Sun*, 22 September 1978.
21 *The Times*, 22 September 1978.
22 *Africa News*, 18 December 1978.
23 *Africa News*, 23 December 1978.
24 *Council on Economic Priorities Newsletter*, 4 December 1978.
25 Interview with Bernard Rivers and Reed Kramer, 1 December 1978.

26 *Council on Economic Priorities Newsletter*, 4 December 1978.
27 Interview with Reed Kramer, 13 December 1978 (published in *Africa News*, 9 March 1979).
28 Interview with Reed Kramer, 14 December 1978.
29 Interview with Reed Kramer, 14 December 1978.
30 Treasury Investigation of Charges made Against the Mobil Oil Corporation, p. 17.
31 UN Sanctions Committee, 10th annual report, p. 267.
32 Letter from J. Marot (Total), 25 November 1977.
33 UN Sanctions Committee, 11th annual report, p. 51.
34 *The Times*, 25 October 1978.
35 Foreword to *A Review of the Bingham Report*.
36 *The Times*, 25 October 1978.
37 Hansard, Commons, 7 November 1978, col. 747.
38 Hansard, Commons, 7 November 1978, col. 750.
39 Hansard, Commons, 7 November 1978, col. 756.
40 Hansard, Lords, 9 November 1978, col. 465.
41 Hansard, Lords, 9 November 1978, col. 465–6.
42 Hansard, Commons, 15 December 1978, col. 1187.
43 Hansard, Commons, 15 December 1978, col. 1187.
44 Hansard, Commons, 15 December 1978, col. 1183.
45 Hansard, Commons, 1 February 1979, col. 1749.
46 *Financial Times*, 13 February 1979.

Part Two

BUSTING SANCTIONS

6 Rebellion

September–December 1965

UDI was solemnly proclaimed on 11 November 1965. Ian Smith announced that Rhodesia's links with Britain had been cut, and that his regime had 'struck a blow for the preservation of justice, civilisation and Christianity'.[1] For the white minority in Rhodesia this was a way of killing any moves towards African rule. But back in Britain Wilson faced what he himself described as 'the most complicated issue which any government of this country has ever had to face in this century'.[2]

It has been claimed that the oil companies actually helped to create the Rhodesian problem, with Shell and the other oil companies being alleged to have encouraged the Rhodesian rebellion through their support to the Smith regime before UDI. It is significant that Wilson himself has recently taken these claims seriously. After the publication of the Bingham Report he told Parliament that a further investigation should focus on the suggestion that Smith was 'pushed into UDI by the oil companies saying that they would look after him and promising "there's nothing to worry about, old chaps" '.[3]

The source of these startling allegations about the involvement of the oil companies is Jardim, who claims in his book *Sanctions Double-Cross*, that Shell pledged that oil supplies to Rhodesia would continue even in the event of a break with Britain. Jardim had visited Salisbury on 14 October 1965, barely a month before UDI, but to his surprise, he found that the Rhodesians were displaying 'the utmost unconcern' about oil sanctions.[4] The explanation, he soon learned, was simple. Just a few days earlier the Rhodesian Secretary for Commerce, D. H. Cummings, had received an important visitor from London. The caller was a senior Shell executive, who had apparently given assurances that an oil embargo 'was out of the question'.[5] Sanctions against Rhodesia, the Shell representative had argued, would be impossible since Zambia, which depended heavily on the Rhodesian Umtali refinery, would also be hit. Three days later Jardim reported to the Portuguese Prime Minister that the Rhodesians 'had received an envoy from Shell in London who had given the most express guarantees as to the improbability of an embargo on the supply of fuels' (B4.5).

Jardim claims that his account was subsequently confirmed by a Portuguese engineer, Pinto Coelho. According to Coelho a civil servant at the Rhodesian Ministry of Trade had revealed that 'one of the directors of Shell from London had recently come to Salisbury; the latter had assured him that come what may Shell would continue to supply the market' (B4.5). Jardim later concluded: 'No doubts remained with us that the Rhodesian officials had absolute faith in the guarantees received from the big international oil companies.'[6]

Jardim's testimony should be considered seriously since it was recorded in contemporary documentation, at a time when he would have had little motivation to fabricate the report. But the major weakness in his evidence is that he has been unable to recall a vitally important detail – the name of the Shell representative from London. As Jardim explained to Bingham: 'They told me the name, but by that time I was not really interested [in] . . . the name of the man . . . I was more interested about the physical problem of transporting products.'[7]

In spite of this flaw, further evidence discovered by Bingham may substantiate Jardim's allegations. He found that a Shell Director had indeed visited Rhodesia on 8 October 1965, only a week before Jardim's arrival. Ironically this was the very day that Wilson first threatened Smith with sanctions during the Rhodesian leader's final visit to London. The Shell Director who had arrived in Salisbury was Dirk de Bruyne, a Dutchman who had just taken over as Shell's Co-ordinator for Africa. De Bruyne subsequently played an important role in the oil sanctions story for many years to come.

The circumstances surrounding de Bruyne's visit were somewhat unusual. A few days before leaving London, he had received a long cable from Peter Jamieson, head of Shell Rhodesia. The message outlined a suggestion from the Rhodesian authorities that if both the British and the Americans imposed sanctions in the event of UDI, then the French company of Total 'might be prepared to fill the breach' (B4.12). Total, it was proposed, could supply crude oil to the Rhodesian refinery as part of a swap arrangement. The British and the American oil companies would then provide matching amounts of crude oil to Total elsewhere.

Jamieson's cable concluded with the news that the Rhodesian Government was hoping to raise 'this difficult subject' during de Bruyne's impending visit. But de Bruyne denies discussing the proposed swap arrangement during his trip, or indeed giving any assurances about future oil supplies. He told Bingham that his reply to the Rhodesians had been blunt: 'I am sorry, that is not a point for discussion here, that sort of thing we can't discuss. We can't say what sort of circumstances will arise' (B4.13). Bingham accepted de Bruyne's explanation, and sanitised the meeting by describing it as 'a courtesy visit' (B4.13). But the politically tense period immediately before UDI was certainly an unusual moment for a courtesy visit from a London Director of Shell. It is also difficult to understand why a man of de Bruyne's experience would have agreed to such a potentially compromising meeting with the Smith regime, merely in order to give a negative response to the Rhodesians.

Further evidence to suggest that the oil companies may well have given some sort of assurances to the Smith regime emerges in a Shell Rhodesia note of March 1969, which claims that at the time of UDI, the international oil companies 'gave the Government supply undertakings which they immediately breached' (B/A 272). This reference adds credence to the suggestion that the Rhodesians *believed* they had assurances that oil would not be cut off in the event of UDI.

Bingham himself admits that he was unable to 'adequately account for the confidence shown by the Rhodesians over oil supplies' (B4.16). He points out that it seemed 'over-whelmingly unlikely' that Shell or BP would have *deliber-ately* encouraged the Rhodesians to embark on UDI, since this 'could scarcely advance the Groups' business interests in Rhodesia but had a great and obvious capacity to injure them'. Nevertheless, he adds, it is possible that de Bruyne's comments were either 'misconstrued' by the Rhodesian Ministry of Commerce or that 'undue weight was given to an unguarded expression of opinion by someone'.

The head offices of Shell and BP would, it is true, have had little interest in actively encouraging UDI for its own sake. But it should be remembered that the senior execu-tives of the oil companies in Southern Africa were all white, and that many of them personally supported the Smith regime. As Bingham discovered, virtually all white South Africans employed by Shell and BP had a 'strong pro-Rhodesian sympathy' (B14.4). It has also been pointed out that at this time Shell was negotiating for concessions in Angola from the Portuguese authorities; it is possible that the company may have thought that 'support for Smith would aid success.'[8]

Action taken by the international oil companies in an attempt to preserve their substantial stake in the Rhodesian market may well have had the effect of encouraging the Smith regime to believe that it would be able to survive an international embargo. As Freitas Cruz, Portugal's Consul-General in Salisbury, later explained: 'The oil companies were eager to inform the Rhodesian authorities that there would be no problems, because they were worried about competitors coming in.'[9] They may also have feared that failure to support the white regime in Rhodesia would have meant a considerable loss of goodwill in South Africa, which then consumed a third of the total oil supplied to the African continent.

*   *   *   *

While the Smith regime was seeking guarantees from the oil companies, the British Government was attempting to en-sure that UDI never actually occurred. Wilson's problem, however, was that his arsenal was bare, since he had already publicly announced that Britain was unwilling even to consider the use of force to stop a rebellion in Rhodesia. It is difficult to explain why Wilson, an experienced negotiator, had thrown away his strongest card before the contest had even begun. If force was not to be considered, economic sanctions remained the only deterrent. Even then, however, it was only at the eleventh hour that the British Prime Minister actually threatened the Rhodesian leader.

On 8 October 1965, barely a month before UDI, Wilson first raised the spectre of sanctions, lamely warning the Rhodesian leader, then on a visit to London, that even if Britain took no action, others might, and 'demands for some form of sanctions would be overwhelming'.[10] But already it was too late; the decision to break with Britain had been made. Rhodesia, Smith believed, could survive an economic war. The only question was *when* UDI would come.

On 12 October 1965, the day Smith left London, Wilson felt that 'there was a general expectation that UDI would come within a very few days'.[11] Certainly the more extreme members of the Rhodesian cabinet were impatient to break with Britain. But UDI was not declared until a month later. The interesting explanation recently offered for the delay is that Smith may have realised that further contingency planning was necessary to protect his country's vital oil supply.

Oil was undoubtedly Rhodesia's weakest point. If the oil supply dried up, then road and air transport would stop, industrial production would slow down to a halt, and the lack of fuel for tractors would hit the modern agricultural sector. Equally serious for the Rhodesian regime, the armed forces would lose their mobility, making Smith's regime vulnerable to internal unrest or external military intervention. Rhodesia's oil stockpile would only have lasted the country a few weeks. Once this supply had been exhausted, the nation would be on the brink of total collapse.

Smith's civil servants carefully analysed the problems that Rhodesia would face in the event of sanctions. Only a few months before UDI Rhodesia's first oil refinery had been opened at the border town of Umtali. This new refinery was fed with crude oil by a pipeline from the Mozambican port of Beira. An international embargo might therefore cut the flow of crude oil to Beira, close the Umtali refinery, starve

114

the marketing companies of fuel, and bring the Rhodesian economy to its knees.

At first sight the fact that Rhodesia was landlocked might have suggested that it would be particularly vulnerable to sanctions. The country, after all, was dependent on two neighbours – Mozambique and South Africa. Mozambique's support was particularly important since Rhodesia's two main rail outlets to the sea ran to the ports of Lourenço Marques and Beira. When UDI actually came, both South Africa and Portuguese-ruled Mozambique remained neutral in the dispute between Rhodesia and Britain. But the white communities in Mozambique and South Africa were naturally sympathetic towards their fellow European settlers in Rhodesia. They also believed that their own survival would be strengthened if the advance of black rule could be halted at Rhodesia's northern border, the Zambezi River. In practice, the 'neutrality' of South Africa and Portuguese-ruled Mozambique therefore meant a refusal to participate in sanctions; and the fact that Rhodesia was landlocked, and therefore had to send its foreign trade through neighbouring countries, ultimately made it easier for the Smith regime to defy the UN embargo.

Fuel was of such importance that Smith dealt with the matter personally. Two days after his return from London he summoned Freitas Cruz, Portugal's trusted Consul-General in Salisbury, to discuss ways of safeguarding Rhodesia's oil supply. Cruz reassured Smith that Portugal would continue to allow Rhodesia to import its petroleum requirements through Mozambique if sanctions were imposed. The Consul-General also agreed to forward a request for several Portuguese experts from the Sonarep refinery in Lourenço Marques to assist the Rhodesian Ministry of Trade in arranging oil imports. Detailed contingency planning, already well in hand just four days after Wilson's first warning of sanctions, was later to prove of vital importance.

At this time Rhodesia had sufficient petrol to last only twenty-seven days. Supplies of certain other oil products, such as aviation fuel, were even lower. Building up the stockpile was thus an urgent priority, and work proceeded quickly: the storage tanks at the Umtali refinery were filled to the brim with 260,000 barrels of crude oil (about a month's requirements for Rhodesia) and the larger tanks at the port of Beira stored further supplies. This stockpile, built up in the precious weeks before UDI, later had a crucial effect in cushioning the country against the initial impact of the embargo.

On 25 October 1965 Wilson flew to Salisbury in a final attempt to prevent UDI. After four days of gruelling bargaining, the two leaders met at a formal dinner. Wilson took Smith aside at the end of the meal to warn him against UDI and for the first time the British Prime Minister threatened sanctions. Oil, he said, would be prevented from reaching Rhodesia, even if it meant blockading Beira. If Smith retaliated by cutting Zambia's supply, then oil would be airlifted to Lusaka.

Next morning Smith denied press leaks that Wilson had threatened an oil embargo. The previous evening, the Rhodesian leader claimed, had been 'a pleasant social gathering which ended in the relation by both sides, and especially the British, of a splendid repertoire of after-dinner stories'.[12] Wilson, who later described the dinner in vivid detail, painted a rather different picture of the proceedings. One of the Rhodesians apparently capped Commonwealth Secretary Arthur Bottomley's Essex cricket story with 'a tale of incredible prurience'.[13] The finale on the Rhodesian side was a performance by Lord Graham, who acted out the story of an American girl who was apparently better at 'displaying her charms' than dancing. Evidently so far as the Rhodesian Cabinet was concerned Wilson's threats of sanctions had fallen on deaf ears. Indeed, the rugger club atmosphere of the dinner suggested that the Rhodesians felt confident enough deliberately to snub the visiting British Prime Minister.

On the last morning before his departure, Wilson told Smith that 'economic war' would follow UDI; sanctions would lead to the disintegration of the Rhodesian economy, and that would spell the downfall of Smith and his replacement by an African leader. In fact, Wilson added, UDI would probably accelerate the transition to majority rule. The British Prime Minister specified that an oil embargo would be imposed. Forty-eight countries would participate, he claimed – a remark which appeared to suggest that careful contingency planning had been taking place behind the scenes.[14] The major oil-producing nations would cut supplies, and the international oil companies would be reluctant to break the ban.

Subsequent events suggest that Wilson was either rather naïve or, more likely, simply bluffing. For at that very moment, the oil companies were building up stockpiles inside Rhodesia. On the very afternoon of Wilson's departure from Salisbury, Freitas Cruz again called on Smith to offer reassurances that Portugal's supportive policy remained unchanged. With the apparent backing from both Portugal and the oil companies, Rhodesia could now afford to make the final break with Britain.

*   *   *   *

UDI finally came on 11 November 1965. Jardim has recounted that on that day he was sitting in the English atmosphere of the Salisbury Club together with the Minister of Commerce, listening to Smith's proclamation on the radio. A few minutes later the Minister told Jardim that the companies of the Umtali refinery, 'represented by Shell/BP', had sent him information that led him 'to believe that a blockade in fuel supplies to Rhodesia was improbable' (B4.7).

Back in Britain the Prime Minister immediately denounced the Rhodesian rebellion to a packed House of Commons, but confirmed that the Government was not considering the use of force. Wilson's reaction angered many African leaders, who noted that Britain had not shirked using force elsewhere in its far-flung empire when rebellions had been led by blacks. Liberal MP Jeremy Thorpe was among those who spoke out for tough action. He told Parliament that 'it would be a fantastic position if this country were incapable of putting down a rebellion of a population the size of that of Portsmouth.'[15] Only twenty years earlier, he argued, the United Kingdom had governed a quarter of the world's population, and at the time of UDI, Britain's military budget was four hundred times greater than that of Rhodesia.

Rhodesia, nevertheless, did have its own security forces, numbering 3,400 men, and a well-equipped air force, and it was the Europeans, not the 'natives' who were in revolt. Some doubts were expressed as to whether British troops

117

would have obeyed orders to put down their own 'kith and kin'. Wilson also faced political problems at home. The Labour Government's majority in the Commons had fallen to only one, and the Conservative Party, although divided on how to deal with Rhodesia, was united in its opposition to military intervention. A serious balance of payments crisis merely increased the Government's reluctance to embark on a risky venture overseas.

Admittedly military intervention was no easy option. Yet as Robert Good, the American Ambassador in Lusaka, later declared, 'it seems reasonable to assume that had the British response been swift and decisive, filling the vacuum of Rhodesian self-doubt and confusion following UDI, resistance would have been minimal and certainly manageable'.[16] Probably Wilson's most serious mistake was to have stated publicly that force would not even be *considered* in the event of UDI.

The Prime Minister told Parliament that the Government had decided to opt for sanctions, and that economic pressure would be brought to bear on the illegal regime to force a return to legality. The first batch of British sanctions was announced on the very day of UDI. British aid and the export of capital were halted; merchandise exports were no longer covered by the Export Credits Guarantee Department; Rhodesia was removed from the Commonwealth Preference Area; and a ban was imposed on the import of Rhodesian tobacco and sugar. But when Jo Grimond, the Liberal leader, asked Wilson about an oil embargo the Prime Minister announced blandly, 'We have no proposals to make on this subject' – a reply which suggested that the Rhodesians' optimism over oil supplies may indeed have been justified.[17] Britain's decision not to impose oil sanctions immediately after UDI gave the Rhodesians a further breathing period to build up stockpiles.

One of the problems was that Britain brought in sanctions on a piecemeal basis, which tended to cushion the shock for Rhodesia. They therefore never amounted to the hard-hitting blow which was needed to topple the rebel regime. Moreover, Wilson's confidence in the sanctions weapon was hardly backed up by the history books. The most celebrated occasion on which they had been introduced had been against Italy in 1935 after Mussolini's invasion of Ethiopia.

More recently, economic warfare had also been used by the Arab states against Israel, by the United States against Cuba and other Communist countries, and by black Africa against South Africa. In all these cases sanctions had had some effect, but they had not by themselves secured major political changes. There was therefore widespread cynicism as to whether sanctions would actually quell the Rhodesian rebellion.

Certainly many of Wilson's Cabinet colleagues were sceptical. Just a week after UDI Richard Crossman predicted: 'the Rhodesians will be holding out successfully in six months' time, and even the oil sanctions the UN are talking about will be ineffective with Portugal and South Africa neutral on the side of Rhodesia'.[18]

Twelve days after UDI Wilson was again asked whether oil sanctions would be introduced. 'There are many technical and economic factors to be examined,' he replied, and an oil embargo is 'bristling with difficulties.'[19] Careful planning was necessary, he said. On 7 December 1965 Michael Foot asked the Prime Minister when these preliminary studies would be completed. Wilson weakly replied that 'this will inevitably take some time'. Two days later Peter Jamieson, head of Shell Rhodesia, arrived on a secret visit for discussions with the Ministry of Power in London. It is interesting that this is the only instance noted in the Bingham Report of Shell or BP executives discussing the likely effect of an oil embargo with the British Government before the Sanctions Order was actually introduced.

Since Smith had been threatening UDI for over a year, it seems incredible that the British Government had undertaken so little planning. The Commonwealth Relations Office had done some contingency planning on sanctions before UDI, and the matter had been examined by a Cabinet Office committee chaired by Deputy Under-Secretary Arthur Snelling. But it seems that relatively little effort was devoted to analysing the effectiveness of an oil embargo considering the importance of the subject. Certainly there is a remarkable contrast between the careful plans which were formulated by the Rhodesians and the oil companies, and the lack of planning undertaken by Britain.

Wilson's comments immediately after UDI on the difficulties of oil sanctions may, it is true, have been a subtle trick to discourage Smith from pursuing his own contingency plans. But if this was his intention, it failed: the Rhodesians continued to make a concerted attempt to increase their stockpiles. Behind Wilson's initial emphasis on the problems of an oil embargo there probably lurked a hope that it would never be necessary actually to use the oil weapon. But the British Prime Minister was soon to realise that he had at least to go through the motions of taking tough action, in order to forestall African demands for the use of force.

* * * *

On 17 November 1965, following mounting pressure at the UN, Foreign Secretary Michael Stewart told the Security Council that Britain was 'ready to envisage the possibility of wider economic measures, including an embargo on oil.' Three days later the Security Council approved a resolution calling on UN members 'to break all economic relations with Southern Rhodesia,' and recommending an embargo on oil and petroleum products. Only France abstained, on the grounds that Rhodesia was a purely British responsibility, and all other Security Council members voted in favour. The French abstention, however, was a foretaste of her unenthusiastic attitude towards the enforcement of sanctions.

The Security Council resolution merely recommended the imposition of sanctions, and was not mandatory. Britain stressed that a careful study would have to be made before it was willing actually to introduce an oil embargo. Within a few days, however, several oil-exporting countries banned the sale of crude oil to Rhodesia. Iran, then the main supplier of the Umtali refinery, cut off shipments to Rhodesia on 22 November. Kuwait, whose national oil company had a small shareholding in the Umtali refinery, and Libya also took similar action. The international oil companies respected these embargoes, but immediately diverted shipments from elsewhere to the Umtali refinery; Abu Dhabi, another Gulf producer, was a convenient source. As a British Protectorate, its external affairs were the UK's responsibility, although oil marketing was technically an internal matter under the control of the Abu Dhabi Government.

Early in December 1965 a BP tanker – the patriotically-named *British Security* – steamed out of Abu Dhabi carrying 80,000 barrels of crude oil for Rhodesia. A BP Managing Director, William Fraser (later Lord Strathalmond), confirmed that the company would fulfill its contract to supply the Umtali refinery 'until the British Government asks us not to.' 'There is no oil embargo,' was the Foreign Office's response.[20] The *British Security* therefore sailed into the port of Beira on 14 December carrying the last shipment of crude oil that reached Rhodesia.

The Government's shareholding in BP ensured that the voyage of the *British Security* whipped up an international storm. Wilson justified the decision to let the *British Security* proceed by pointing out that two other tankers were 'stacked' behind the BP vessel. 'Interference with the *British Security*,' he explained, 'would not make any difference with the two others immediately behind it.[21] There were enterprising businessmen who could divert tankers in mid-ocean to Beira. 'The oil trade contains a lot of privateers,' he added.[22] Later events suggested that these pirates included Shell and BP themselves.

Even after UDI the Rhodesians seem to have been well-served by the oil companies. The storage tanks at the Umtali refinery had already been filled to the brim; the Shell/BP, Mobil, and Caltex tanks at the port of Beira were full; a tanker was standing by in the harbour ready to discharge further crude oil supplies; and on 18 November Mobil asked Mozambique Railways for special facilities to carry additional refined products from Lourenço Marques into Rhodesia. At the end of November, Smith received the encouraging news that in just over a month, Rhodesia's stockpile had been increased from twenty-seven days' to three months' supply. Bingham acknowledged that stockpiles in Rhodesia had been low in September and then rose in October and November, and added, 'It seems very likely that this increase was deliberately achieved' (B4.23).

Building up stocks in Rhodesia was doubly beneficial to the Smith regime since it both strengthened Rhodesia and weakened Zambia. At the time of UDI, Zambia also obtained most of its oil supplies from the Umtali refinery, which meant that as stockpiles were built up in Rhodesia, there was less oil available to send on to neighbouring Zambia. This effect was confirmed by figures showing the level of stocks in the two countries. On 16 October Zambia had twenty-three days' supply of oil – virtually the same as Rhodesia. But by 11 December, stocks in Zambia had fallen to only eleven days.

According to Bingham, it had not been possible to determine whether the oil companies had deliberately attempted to run down stocks in Zambia. But it is clear that Zambian stocks were reduced, whether intentionally or not, and the effect was certainly to make Britain hesitate before imposing oil sanctions against Rhodesia (B4.28).

Zambia's vulnerability was increased still further because the Rhodesian Government regularly received full details of the level of stocks held by its northern neighbour. The oil companies provided the Ministry of Commerce in Salisbury with a breakdown of stocks, intake, and consumption of different oil products in Zambia. This information on the imbalance of stocks in Rhodesia and Zambia bolstered the confidence of the Smith regime: Zambia was 'Rhodesia's hostage'.

Initially Wilson had suggested that sanctions should be 'quick and sharp' to avoid 'a long drawn-out agony in Rhodesia'.[23] Yet Britain's actual reaction to UDI seemed to suggest that the UK was unwilling to take tough action to quell the rebellion. Black African nations therefore never believed that sanctions would topple the Smith regime.

On 5 December 1965, the Organisation of African Unity (OAU) met in Ghana to consider the Rhodesian crisis, and the OAU Council of Ministers called on Britain to take decisive action – including an oil embargo – to crush the rebellion, threatening to break off diplomatic relations with Britain if Smith was not toppled within ten days. After the OAU meeting, a Ghanaian Minister led a demonstration outside BP's office in Accra. The Minister, in an angry mood, called for a consumer boycott of BP petrol, and warned that similar campaigns would be mounted in other African states.

If Wilson had immediately announced an oil embargo, many of the African states which eventually broke off diplomatic relations with the UK would probably have reconsidered their decision. But ten days passed after the OAU resolution. Thirteen African countries, including two Commonwealth members, then took the unprecedented step of cutting their diplomatic ties with the UK. On 16 December 1965, the very day that the OAU deadline expired, Wilson spoke at the United Nations in New York. As he mounted the rostrum to address the General Assembly, some one hundred delegates from a score of African countries ostentatiously left the hall in a silent protest against Britain's unwillingness to take effective action against the Rhodesian rebellion. Britain's relations with black Africa had reached their lowest point.

NOTES

1 Robert Good, *UDI*, (Faber & Faber, 1973), p. 16.
2 Elaine Windrich, *Britain and the Politics of Rhodesian Independence*, (Croom Helm, 1978), p. 11.
3 Hansard, Commons, 7 November 1978, col. 754.
4 Jardim, *Sanctions Double-Cross*, p. 17.
5 Jardim, *Sanctions Double-Cross*, p. 17.
6 Jardim, *Sanctions Double-Cross*, p. 18.
7 Interview, 25 May 1978.
8 Memorandum on Sanctions Violations, submitted by the United African National Council to US Senate, 10 March 1977, p. 9.
9 Interview, 2 September 1978.
10 Kenneth Young, *Rhodesia and Independence: A Study in British Colonial Policy*, (J. H. Dent, 1969), p. 236.
11 Harold Wilson, *The Labour Government 1964–70*, (Penguin, 1974), p. 199.
12 Kenneth Young, *Rhodesia and Independence*, p. 262.
13 Harold Wilson, *The Labour Government 1964–70*, (Penguin, 1974), p. 215.
14 Letter from Rowland to Trade Secretary Edmund Dell, 7 January 1977, p. 5.
15 Robert Good, *UDI*, p. 56.
16 Robert Good, *UDI*, p. 58.
17 Hansard, Commons, 11 November 1965, col. 360.
18 Richard Crossman, *The Diaries of a Cabinet Minister*, vol. i, (Hamilton & Cape, 1975), p. 382.
19 Hansard, Commons, 23 November 1965, col. 249 & 252.
20 *Sunday Times*, 3 December 1965.
21 Hansard, Commons, 7 December 1965, col. 243.
22 Hansard, Commons, 1 December 1965, col. 1437.
23 Elaine Windrich, *Britain and the Politics of Rhodesian Independence*, p. 69.

7   Weeks Not Months

December 1965–March 1966

After the humiliation of the African walk-out at the United Nations, on 16 December 1965, Wilson's next stop was Washington. Just a few hours after facing a hostile United Nations, the British Prime Minister was sitting in the elegance of the White House. Britain had already decided to introduce an oil embargo. But as Wilson later admitted, sanctions would only be effective if they were 'backed by other countries with big oil interests'.[1] Mobil and Caltex together had a forty per cent share of the Rhodesian market, and the British Prime Minister therefore needed US support for his next move against the Smith regime.

Initially there had been some fears that Washington would refuse to cooperate. A few days before, Deputy Prime Minister George Brown had reported to Cabinet that the Americans were being completely negative over sanctions. They had, he said, responded by asking when the first British battalion would be arriving in Vietnam.[2] But Wilson's personal plea to the US President in Washington succeeded. The British Prime Minister quickly convinced Lyndon Johnson that tougher measures were needed against the rebel regime, and that otherwise there would be very strong African pressure for the use of force. America agreed to back the Rhodesian oil embargo, in return for British political support for US policy towards Vietnam. President Johnson, a US diplomat commented, could now count on Wilson 'to curb the most extreme demands of Labour backbenchers who were by now openly opposed to US policies in Vietnam'.[3] It was now clear that Britain's decision to back American intervention in Vietnam was therefore partly determined by Wilson's need for US support over Rhodesian sanctions.

124

News of American support for oil sanctions was immediately flashed back to London. At 6.30 pm on Friday 17 December 1965, George Brown announced that the Queen had just signed an Order in Council introducing an oil embargo against rebel Rhodesia. 'Her Majesty, in exercising the powers conferred on Her by the Southern Rhodesia Act 1965, is pleased, by and with the advice of Her Privy Council, to order ... , ' it began grandly, going on to make it illegal both for companies registered in Britain and for British citizens to 'do any act calculated to promote the supply or delivery of petroleum' to Rhodesia.[4]

But the crucial flaw in the UK Sanctions Order, which should have been obvious from the very start, is that it only covered companies registered in Britain, not their subsidiaries which were incorporated in foreign countries. If *every* government in the world (with the exception of Rhodesia) had introduced – *and* enforced – similar legislation, then sanctions would probably have worked. But right from the start it should have been clear that South Africa, in particular, would be unlikely to accept sanctions, and that this would enable the South African subsidiaries of Shell and BP (and the other international oil companies) to supply Rhodesia without breaking any law. This was soon to prove the complete undoing of sanctions.

* * * *

The oil embargo marked the end of the bipartisan approach to the Rhodesian crisis at Westminster. Unlike the earlier measures, the new Sanctions Order did not pass through Parliament without a division. The Conservative Party was split three ways: thirty-one Tories supported the Government's decision to introduce oil sanctions; forty-eight voted against, on the grounds that punitive action should not be taken against Rhodesia; and the rest of the Conservatives, including the Front Bench, abstained. This division in Westminster encouraged Smith's belief that Rhodesia would survive sanctions – especially as the Labour Government's tiny minority over the opposition parties meant a general election could not be long postponed.

Wilson now concentrated on his most urgent task: to cut the flow of crude oil to the Rhodesian refinery. Systematically he worked on each link in the complex supply chain leading to Umtali. Since there was no refinery at Beira, any crude oil which arrived at the port was clearly intended for Rhodesia; the first step, therefore, was to ensure that no further deliveries were made.

On the day that the Sanctions Order was introduced, two oil tankers were already on the high seas bound for Beira. Shell's *Staberg* was approaching the port bearing 80,000 barrels of crude oil for Rhodesia. Shell Centre in London immediately ordered the ship to alter course. Following in the *Staberg*'s wake was the *Tamarita*, a Norwegian tanker chartered to Aminoil (American Independent Oil Company). After the US Government had announced its support for the oil embargo, Aminoil ordered the *Tamarita* to discharge at the Kenyan port of Mombasa.

All the oil companies with shareholdings in the Umtali refinery made it clear that they would cut off supplies of crude oil. But there was always the danger that the Rhodesians would try to purchase shipments from brokers and other intermediaries. As Wilson had warned Parliament, just three days after the introduction of the Sanctions Order: 'I gather that it is possible for spivs to invade the oil trade for the purpose of defeating the laws that have been made.'[5]

The second link in the supply chain therefore had to be cut: the 184-mile pipeline which runs from Beira to the Umtali refinery. Lonrho, which had taken the initiative to build the project, held a sixty-three per cent shareholding in the Companhia do Pipeline Mocambique-Rodesia (CPMR), and the remaining shares were held by the Portuguese Dias da Cunha family. But despite its majority shareholding, Lonrho did not actually control the company. Because the pipeline ran through Portuguese-ruled Mozambique, the company was registered in Lisbon, and the Dias da Cunha family was given an equal voting share. Mozambique Railways also had the right to nominate one member to the Board. Altogether there were therefore five Portuguese and four British Directors running the pipeline company.

The CPMR Board met in Lisbon four days after the introduction of the Sanctions Order. The fact that one of the British Directors on the pipeline board was Angus Ogilvy gave the situation an added delicacy. Ogilvy's royal connections would have made it extremely embarrassing for the pipeline company to have pumped oil to a regime in rebellion against the Crown. Rowland, the senior Lonrho representative, worked behind the scenes to persuade the Portuguese Government that it might endanger Portugal's own oil supply if the country was seen as the main culprit over the failure of sanctions. The Lisbon authorities therefore discreetly suggested to the Mozambique Railways representative that it would be better if he remained neutral and did not attend the controversial board meeting. The Portuguese Chairman of the pipeline company also decided not to use his casting vote. On 21 December 1965 the Companhia do Pipeline Mocambique-Rodesia therefore decided by four votes to three that the pipeline would close after existing stocks had been pumped to Umtali.

For the present, at least, the pipeline would not handle Rhodesian oil. But the British Government was uncomfortably aware that this decision could easily be reversed by the built-in Portuguese majority on the company's board. A special arrangement was therefore made to compensate Lonrho for the loss of revenue that it suffered through the closure of the pipeline. From April 1966 Lonrho received £54,000 per month to help finance the costs of maintaining the pipeline. It was a unique arrangement, since Lonrho was the only company ever to receive compensation for losses suffered as a result of Rhodesian sanctions. But the pipeline company held enormous strategic power, and it was essential that the British Directors should be encouraged to do all they could to prevent the pumping of oil into Rhodesia. (These compensation payments were subsequently halted by the British Government in September 1966.)

The final link in the supply chain was the Umtali refinery itself, owned by Central African Petroleum Refineries (Capref), a consortium of international oil companies. Shell, BP, Mobil, and Caltex are the major shareholders, each with a seventeen and a half per cent stake, and the remaining shares are held by Total, Aminoil, and the Kuwait National Oil Corporation. Capref, however, was registered in Salisbury, and after UDI it fell – like the local marketing firms – under the authority of the Rhodesian regime. The international oil companies which own Capref therefore lost all formal control over its Rhodesian operations, and there

127

was nothing that Britain could do to prevent the refinery from processing any crude oil that might reach Umtali.

By January 1966, however, the British Prime Minister was confident that the oil embargo would be the final blow which would topple the Smith regime. No crude oil had been delivered at Beira since the introduction of the Sanctions Order. The storage tanks at the port had run dry by the end of December 1965, and pumping through the pipeline had then ceased. Supplies of crude oil stored at the Umtali refinery were quickly exhausted, and on 15 January 1966 the plant finally closed down. The giant flame from the flare tower at the refinery – an Umtali landmark at night – was extinguished. Petrol rationing had been introduced. Rhodesia, the British Prime Minister believed, was rapidly grinding to a halt.

*   *   *   *

Wilson now confidently claimed that the rebel regime would be toppled 'within weeks, rather than months'. The occasion for this immortal prediction was a meeting of Commonwealth Prime Ministers in the Nigerian capital, which had been called to discuss the Rhodesian crisis. At the summit Britain came under very strong pressure to use force against Smith, but agreement between the United Kingdom and the rest of the Commonwealth looked virtually impossible. However, as Wilson has recounted, a last-minute compromise was finally reached on 12 January 1966.

> By tea time we were further from an agreement than ever . . . An hour later we were back, fully agreed – on our terms – subject to a short drafting session. But this was partly achieved by my phrase 'weeks not months', based on advice we were receiving that the oil sanctions and the closure of the Beira pipeline would bring the Rhodesian economy to a halt. We had good reason to believe that Portugal would not challenge the determination of the UN, nor seek to encourage sanction-breaking. We were misled, but what I said to my colleagues appeared at the time to be a safe prophecy.[6]

Commonwealth leaders, who were by no means as convinced as the British Prime Minister, insisted that Wilson's phrase should be placed on-the-record in the conference's final statement. The Lagos Communiqué therefore 'noted the statement by the British Prime Minister that on the expert advice available to him the cumulative effects of the economic and financial sanctions might well bring the rebellion to an end within a matter of weeks rather than months'.

Only two explanations are possible for Wilson's confident prediction. It may have simply been a bluff to maintain lukewarm Commonwealth support for sanctions. But if this was Wilson's intention, it was remarkably shortsighted, since it would inevitably be exposed within several weeks. It appears, then, that Wilson may be correct in claiming that he had been given misleading information by his advisors. 'One always regrets being wrong,' he later admitted. 'It was wrong, yes, but it was not wrong on the basis of the evidence available at that time.'[7]

Rhodesia's oil stocks, it is true, *would* only have lasted a matter of weeks: when the Sanctions Order was introduced its stockpile would have amounted to about three months' supply, and by the time of Wilson's prediction this might have been further reduced to just over two months. Although consumption had been reduced by rationing, and existing stocks could therefore have been stretched out to last for a slightly longer period, Rhodesia clearly only had enough oil to survive for a matter of weeks.

But the flaw in Wilson's prediction was that it was based on the assumption that *no* further oil would reach Rhodesia. In fact, few political observers believed that South Africa or Portuguese-ruled Mozambique would respect the embargo, and sure enough, it was only a few days after Commonwealth leaders departed from the Lagos summit that the first shipments of refined oil were sent from South Africa into Rhodesia. Once this flow of refined oil began from Rhodesia's neighbours, Wilson's whole policy was torn to shreds.

It is a frightening comment on the Government's intelligence-gathering that the British Prime Minister was given such poor advice on the most important foreign crisis that he had to deal with during his period in office. The débâcle over the 'weeks not months' prediction has since

9

come to symbolise the blunders and deceit which were the hallmarks of Britain's Rhodesian policy.

*   *   *   *

It is still unclear exactly who it was who told Wilson that the oil embargo would topple Smith. It is rather easier to say who it was *not*. The Commonwealth Relations Office, which was the main Government department responsible for sanctions, was apparently not responsible for predicting the rapid collapse of the Smith regime. Joe Garner, who was Permanent Under-Secretary at the time, has said that hopes that South Africa would confine exports to Rhodesia to 'traditional' supplies, and that Portugal would respect the embargo 'were not based on any firm evidence available to the CRO'.[8] He added that his staff 'were completely taken by surprise when the [Prime Minister's] statement was made'. Commonwealth Secretary Arthur Bottomley has also admitted that he personally was never consulted about the 'weeks not months' prediction.[9]

Michael Stewart, who was then Foreign Secretary has said that right from the start it seemed 'over-optimistic' to believe that sanctions would take effect so quickly. Wilson's optimism 'certainly did not originate from the Foreign Office', he has claimed, and he always believed that 'it was clear that South Africa and Portugal would not accept oil sanctions'.[10] This account has been confirmed by a key civil servant who was dealing with the issue: 'Wilson's prediction about the downfall of the Smith regime certainly did not come from the Foreign Office.' Finally there has been no suggestion that the Ministry of Power ever expected the oil embargo to bring about the collapse of the Rhodesian rebellion. Indeed the Ministry of Power, partly because of its close links with the oil companies, was always very sceptical about the effectiveness of oil sanctions.

It is much more difficult to say who *did* advise the Prime Minister. Chapman Pincher has suggested that Wilson based his statement 'on a brief provided by the Secret Intelligence Service, also called MI6'. Pincher adds that Wilson later 'manfully resisted what must have been extreme temptation to explain his prediction'.[11] Unfortunately little evidence is available on the role of MI6 in the sanctions story. But it does appear that information from intelligence sources proved notoriously unreliable on this vital aspect of Britain's Rhodesian policy.

Richard Hall, an expert on Central Africa, has claimed that the 'weeks not months' prediction came from 'experts at the Department of Economic Affairs'.[12] At this time William Nield, a senior civil servant at the Department of Economic Affairs, had chaired a committee of civil servants dealing with sanctions. Nield was also present with Wilson at the Commonwealth summit conference in Nigeria.

The *New Statesman* has recently reported that the source of Wilson's optimism was Oliver Wright, then the Prime Minister's Private Secretary at Number Ten responsible for foreign affairs.[13] Wright had apparently just come back from a secret visit to Rhodesia on behalf of the Prime Minister. On his return he apparently told Wilson that the 'wall-papering was good in Salisbury, but cracks are fast beginning to appear, and sanctions would soon bite hard'.

It is therefore still unclear exactly who told Wilson that Smith would be suing for peace 'within weeks'. But it does appear that this particular piece of advice probably did not originate from the three Ministries most intimately concerned with the oil embargo – the Commonwealth Relations Office, the Foreign Office, or the Ministry of Power. Nevertheless, in view of the fact that officials and Ministers from all these departments were privately surprised by the Prime Minister's announcement, it is not clear why they apparently failed to pursue the issue with Wilson. Why did they not ask whether new information had emerged to suggest that the oil embargo would be effective so quickly, or alternatively, *correct* the Prime Minister on this vital point?

The whole incident seems to have been surrounded by self-delusion. Wilson listened to those advisors who were telling him what he wanted to hear, rather than taking advice from the Government departments most directly involved. Even then, however, the Ministries dealing with the oil embargo seem to have been very lax in not questioning the Prime Minister after his public commitment. Once the 'weeks not months' statement had been made, the British Government as a whole became over-confident that South Africa and Portugal would both respect the embargo. But too little effort was made to ensure that the prediction actually came true.

✻   ✻   ✻   ✻

By January 1966 the immediate goal had been attained – to cut the flow of *crude* oil. But if oil sanctions were to be effective, then it was essential that *refined* oil products did not replace the crude oil. Two of Rhodesia's neighbours, South Africa and Portuguese-ruled Mozambique, were obviously sympathetic to the rebel regime, and there was always a strong possibility that the Smith regime would import its oil requirements in refined form through these two countries.

'Right from the start,' the US Ambassador in Zambia has explained, 'the American Government was aware that oil would flow through South Africa and Mozambique – and Washington warned London of this before sanctions were ever imposed.'[14] Indeed, at the end of November 1965, a joint British-American working committee had analysed the likely impact of an oil embargo and apparently decided that it would be rendered ineffective by South African and Portuguese suport for Rhodesia. Nevertheless, the joint working committee decided, it might still be worthwhile 'for whatever marginal effects it would have on the Rhodesian regime and as a diplomatic device with the Africans'.[15]

South Africa's attitude was obviously key. Public opinion among the white community in South Africa backed UDI. The Rhodesian whites, it was believed, were making a valiant attempt to preserve 'civilised standards' against the onrush of black rule. And, after all, the South Africans could not afford to sit back and watch sanctions bring down a neighbouring white regime, knowing that success might well be a powerful precedent for applying similar measures against themselves.

Technically it would not have been difficult for South Africa to provide Rhodesia's oil. South Africa had a well-developed oil industry and three major refineries. Most importantly, however, Rhodesia's oil requirements only represented seven per cent of those of South Africa, so it would be relatively easy for South Africa quietly to pass on the fuel needed to sustain its northern neighbour. The South

Africans realised that the British Government would be very reluctant to push the matter to a confrontation, because of the close economic links between the two countries. A deterioration in British-South Africa relations, it was feared in London, might have a serious impact on the UK economy.

Initially, it is true, the South African Government gave the impression that it might at least partially respect sanctions. On 23 December 1965 the oil companies operating in South Africa were summoned to the Ministry of Commerce. The head of the Ministry opened by saying that the meeting was being held at the request of Prime Minister Dr Verwoerd. The oil companies, Shell recounted, 'were virtually told to consider what he had to say as being the Prime Minister's personal instructions to us'. South Africa, the Secretary for Commerce explained, was determined 'to maintain absolute neutrality in the dispute between Rhodesia and the UK' (B5.7). The South Africans were not willing to impose oil sanctions against Rhodesia. But neither would they endanger their own supplies by taking positive steps to increase oil exports to Rhodesia. The oil companies in South Africa were therefore requested to restrict their trade with Rhodesia to its 'normal' level.

The Secretary for Commerce then went on to explain what this policy meant in practice. The oil companies should continue to export 'traditional' supplies – largely specialised products such as lubricants and aviation gasoline – which had previously been sold from South Africa to Rhodesia. But they were not to meet any orders for other products, such as petrol and diesel (which had normally been supplied from Mozambique). 'Traditional' exports from South Africa represented just under ten per cent of Rhodesia's total oil requirements, and if no petrol or diesel had been sold by South Africa then sanctions might still have been effective. Nevertheless the continuance of 'traditional' supplies represented the first breach of the embargo. It was the thin end of the wedge.

On the day after the meeting with the South African Ministry of Commerce, Shell South Africa contacted London. Dirk de Bruyne, Shell Co-ordinator for Africa, was informed that the company's South African subsidiary was 'resuming "traditional" supplies (only) to Rhodesia ex Durban' (B5.7). It was perhaps surprising that he did not immediately cable back ordering Shell South Africa to keep to his original instructions to end all sales to the rebel regime. Instead de Bruyne merely requested the company to inform him what 'traditional supplies' were involved, and asked Shell South Africa not to increase this trade.

The lawyers at the head offices of Shell and BP in London were now worried. The companies were concerned that they might be committing a breach of British law if they sold any oil to South Africa in the knowledge that a small part of those supplies would end up in Rhodesia. On 31 December John Berkin, the Shell Managing Director responsible for Africa, wrote to the Ministry of Power for advice about the legal implications of continuing 'traditional' exports from South Africa to Rhodesia. He concluded by asking the Permanent Secretary for 'an urgent expression of your views, or, alternatively, a meeting with the Minister early next week' (B5.17).

In the event, however, Berkin did not give an appointment with the Minister, nor with the Permanent Secretary; instead he was seen by an official two levels down – the head of the Petroleum Division (John Beckett). The fact that the meeting was held at a comparatively junior level may well have given the oil companies the impression that the oil embargo had already been assigned a relatively low priority by the Government. The meeting itself was somewhat inconclusive, however, and the oil companies therefore decided to continue supplying their South African subsidiaries.

Initially the South Africans had been worried. There was always the possibility that their own oil supplies might be cut off if they defied the Rhodesian embargo. But the UK Government's apparent decision not to take any immediate action to prevent the British-owned oil companies in South Africa from continuing 'traditional' trade with Rhodesia increased their confidence. Gradually the South Africans realised that they could afford to become more brazen in their defiance of sanctions.

On 25 January 1966 Dr Verwoerd told the South African Parliament that his Government's policy over Rhodesia was one of non-interference. 'If there are producers or traders,' he said, 'who have oil or petrol to sell, whether to this country or to the Portuguese, Basutoland [since renamed Botswana], Rhodesia or Zambia, then it is their business and we do not interfere. We do not prevent them from selling. Because, Sir, if we tried to prevent them we would then be participating in a boycott.'[16] This was a clear sign that South Africa was on the verge of expanding its 'traditional' exports.

The British Government continued to display a remarkable degree of naïvety over South Africa. On 27 January, two days after Dr Verwoerd's statement, Ian Lloyd MP asked in the House of Commons whether the South African subsidiaries of Shell and BP would not 'be bound to take their instructions from Cape Town rather than Whitehall?' Wilson's bland response to this very pertinent question was that 'there were rather broader considerations to be taken into account than that in the particular week in which oil sanctions were imposed'.[17] This appeared to contradict his earlier statements that the introduction of the oil embargo had been delayed until international support had been obtained and the practical difficulties resolved.

The Prime Minister went on to tell Parliament that the difficulties over oil sanctions had been 'much less, thanks to the attitude of South Africa'.[18] On the very same day, however, out in Pretoria the oil companies were summoned for another meeting with the Secretary of Commerce, at which a crucial change of policy was announced. Shell South Africa reported to Shell Centre in London that the company had now been asked that 'we should continue to supply Rhodesia and ... if any local trader wished to purchase from us we should supply him even though we knew he was on-selling to Rhodesia' (B5.27). This meant that the oil companies would now have to begin selling 'non-traditional' oil products, such as petrol and diesel, to Rhodesia or to South African intermediaries who were exporting to Rhodesia.

Shell's head office apparently greeted this news with some dismay. Dirk de Bruyne explained: 'What we had suspected was now happening. We were going to be put under pressure from both sides and the whole conflict was developing' (B5.27). The policies of South Africa and Britain were diametrically opposed, de Bruyne realised, and the oil companies would have to play a delicate balancing game.

On 28 January the Ministry of Power made a belated attempt to take a strong stand. Shell and BP were told that

135

the Sanctions Order would be strictly interpreted over the question of 'traditional' supplies:

> Your company supplies crude oil and other forms of petroleum to your affiliated companies in South Africa and ... a small proportion of those supplies have in the past been used to make onward supplies to Southern Rhodesia. Should such supplies to Southern Rhodesia be made in the present circumstances, then it would appear to Her Majesty's Government that the foregoing provisions of the Order-in-Council will have been infringed (B5.24).

It was a tough stand: the British oil companies were being told that if they sold any oil to South Africa and some of these products were then supplied to Rhodesia, the head offices of Shell and BP would be contravening the Sanctions Order. But unfortunately the Ministry of Power's policy was never actually implemented. The British oil companies were never prosecuted for supplying South Africa, the source of Rhodesia's oil.

The flow of petrol from South Africa to Rhodesia began within a matter of hours of Dr Verwoerd's 25 January announcement. It started as a trickle of 'gifts', organised by various 'Friends of Rhodesia' groups which had suddenly sprung up throughout South Africa. These organisations would hire a road tanker or truck, festoon it with Rhodesian flags and 'Oil for Rhodesia' slogans, and drive up to a service station in the Transvaal to take on supplies. The road tankers would then drive over the Beit Bridge border to be received by local dignitaries in a Rhodesian town.

What began as a somewhat haphazard and freelance method quickly attracted racketeers. Within days however, the operation turned into a regular commercial venture. Early in February, the *Rand Daily Mail* sent a reporter into the northern Transvaal to check feeder roads leading to the Rhodesian border. Here it quickly became obvious that a shuttle service had been organised. On 5 February the *Rand Daily Mail* published a photograph of a road tanker carrying fuel to Rhodesia. On the side of the truck, faintly visible through a thin coat of grey paint, was a large 'P', part of the BP insignia. Robert Good, the US Ambassador in Lusaka, later commented: 'BP was first into the breach, to be followed by Shell and subsequently by other international oil companies . . . London's failure to use effective pressure against BP at the very outset carried far-reaching consequences.'[19]

Soon road tankers with the livery of Shell, Mobil, Caltex, and Total were spotted on the Beit Bridge run. These appeared to belong to the Rhodesian, rather than the South African subsidiaries of the oil companies. Even at this early stage, however, there were probably close contacts between the Rhodesian and the South African firms. Caltex Rhodesia, for instance, borrowed a District Manager from the Johannesburg office of Caltex South Africa who knew the Transvaal intimately and, speaking Afrikaans, was able to tap every possible source of supply in the area (B5.79).

Another sign that the procedures for importing oil into Rhodesia were being regularised was the creation of Genta, which was probably set up in January although not formally incorporated until 16 February 1966. Nominally it was a private company, but it was actually controlled by the Rhodesian government, its main role being to act as an intermediary between the oil marketing companies in Rhodesia and their Southern African suppliers.

Right from the start there were disturbing signs that the oil companies would be reluctant to cut off supplies to Rhodesia. Most of the executives within the oil industry in South Africa were whites who were sympathetic to the Smith regime. The problems caused by these attitudes were described by Louis Walker, head of Shell South Africa, when he first instructed the depot managers not to sell to any traders with Rhodesian licence plates:

> I told them they were not to sell to unusual buyers, strangers and so on. Of course they assured me that they wouldn't but one has to face the reality as to whose side they were on. They were on the side of the Rhodesians. I just went on repeating these warnings and instructions but with very little hope that they were really being observed (B5.89).

The local subsidiaries of the other oil companies appeared equally reluctant to lose their Rhodesian market. Jean Flamand, the former General Manager of Total South

137

Africa, flew out from Paris to instruct his successor, Louis
Deny. According to Jardim, who was present at the meeting
on 29 December 1965, Total South Africa was unwilling to
accept head office's orders to halt Rhodesian sales:

> M. Flamand asked [Deny] if he was prepared to dis-
> obey his instructions from Paris. M. Deny replied in
> the affirmative and this appeared to cause the greatest
> concern to M. Flamand who, in strong terms, sought
> to persuade him to give up this attitude (B5.37).

Deny kept his word, and defied instructions from Paris to
cut off shipments to Rhodesia. A few months later, however,
he was personally to supervise the first direct rail consignment of oil to Rhodesia by one of the international oil
companies.

Even officials from the head offices of Shell and BP
sometimes appeared to be condoning trade with Rhodesia
during the early days of the embargo. Hugh Feetham, a
London-based Shell official responsible for Southern Africa,
told local staff at Lourenço Marques on 9 January 1966 that
'they would have to watch competitors' actions closely, as
any company getting supplies would get the trade and
Jamieson [head of Shell Rhodesia] would have to use his
own judgement if it appeared to him likely that, in the event
of his refusal to purchase supplies from outside Rhodesia,
any company would agree to do so' (B5.98). This was, Bingham admitted, 'at least a veiled invitation to evade sanctions'. Louis Walker, head of Shell South Africa explained the incident by telling Bingham that Feetham 'had
been a salesman all his life and he was undoubtedly very
concerned about the Shell Company of Rhodesia's market
share'.

*   *   *   *

Initially the British Government discounted newspaper
reports that substantial quantities of oil were being sent by
road into Rhodesia. On 11 February 1966 *The Times* quoted
a Commonwealth Relations Office spokesman who predicted
that sanctions would reduce Rhodesia to submission by late
March or early April.[20] The following day, however, British

Embassy officials from Pretoria mounted a round-the-clock surveillance at the Beit Bridge border from a parked car a few yards from the frontier port. Naturally this excited the displeasure of the South African authorities. But Arthur Bottomley, then the Commonwealth Secretary, later recalled that this surveillance showed that 'substantial quantities were *not* going through' to Rhodesia.[21] It now appears, however, that the Government's intelligence-gathering services were simply not doing their job.

Government spokesmen repeatedly suggested that reports of heavy oil traffic at Beit Bridge were being deliberately inspired in Salisbury to give the impression that the embargo had already failed. On 19 February, for instance, the Foreign Office described stories of 1,000 barrels a day of oil being sent to Rhodesia as 'very clearly highly coloured'.[22] Bingham, however, later confirmed that during the month of February 1966 the average quantity of oil sent from South Africa to Rhodesia had indeed been 1,000 barrels a day (B5.85). The road traffic quickly increased, reaching 3,000 barrels a day by May 1966. After that, however, it declined as the rail route from Mozambique came into use.

The British Government could hardly afford to watch South Africa drive a wedge through its embargo, and on 19 February Wilson summoned the South African Ambassador in London, Dr Carel de Wet, to seek an explanation. But this had little effect. In a speech during the South African election campaign the following month, Dr Verwoerd turned to what had become the crucial question – the definition of 'normal' trade – and made it absolutely clear that South Africa would continue to do business with Rhodesia. The South African Prime Minister regretted, he said, that his earlier comments had been misinterpreted. 'Normal' trade did not imply the mere selling of commodities or quantities which had been sold before UDI: in competition everyone tries to sell whatever he can and as much as he can without any brakes being applied. 'In doing so it often happens that one trader gains advantage over another,' the South African Prime Minister explained. 'That does not make it abnormal trade. It only makes it better trade' (B5.11).

The South African Government had cleverly defeated the oil embargo through what now appears to have been a carefully planned strategy. Initially the Government had announced its neutrality: only small quantities of 'traditional' exports of specialised oil products would be permitted. Later non-commercial organisations sent 'gifts' of 'non-traditional' supplies. Once this traffic had been established, private companies quietly joined the trade. Finally, just over two months after the UK Sanctions Order had been introduced, the South African Prime Minister publicly announced that companies in the Republic should seek 'better trade' with Rhodesia. This announcement came at a critical moment, since Rhodesia's oil stocks needed building up. Britain's weak stand over sanctions had thus given South Africa the confidence to defy the oil embargo.

The South African road route via Beit Bridge quickly proved an expensive way of transporting oil. The Rhodesian's therefore turned their attention to the railway that runs from the Mozambican port of Lourenço Marques into Rhodesia. This route offered considerable savings, since railway transport is much cheaper for bulk shipments over long distances. There was also an added advantage in dispatching oil from Lourenço Marques: the local Sonarep refinery was controlled by Portuguese interests, and therefore was not under quite the same restraints that the British and American-owned refineries in South Africa faced in trading with Rhodesia.

The major stake in the Sonarep refinery belonged to Manuel Boullosa, a Portuguese businessman, who held three-quarters of the shares. The remaining shareholding belonged to the French company of Total, which also had a long-term contract to supply its crude oil requirements. At the time of UDI most of Sonarep's output of 19,000 barrels a day was sold inside Mozambique and in South Africa. Consequently there was only a relatively small excess production to be exported to the Smith regime. But Jardim, the Sonarep Manager, did what he could to take advantage of the situation by expanding sales to the new Rhodesian market.

Oil products began to be railed from the port of Lourenço Marques into Rhodesia in February 1966. Complicated procedures were adopted to disguise this trade. Jardim has recalled how special orders for existing customers inside Mozambique were 'fabricated by trustworthy personnel'.[23] These supplies never reached their stated destinations but instead were quietly sent across the border into Rhodesia. To begin with, at least, the quantities involved in this clandestine traffic were small. But soon the Sonarep refinery instructed its South African marketing company to place extra orders. With the cooperation of Mozambique Railways, the tank-cars carrying oil for Sonarep South Africa were then re-routed along the line and diverted to Rhodesia. As Jardim later admitted, 'we were lucky that no derailments or fires ever occurred, as they would have broken our careful cover'.[24]

The British Government seems to have been as surprised about this flow as it had been over the road traffic from South Africa. But again it should have been clear that Portugal was hostile towards sanctions. As Bingham pointed out: 'From a date well before UDI, the Portuguese policy was one of close collaboration with the Smith Government in Rhodesia. After UDI it did all it safely could to ensure continued supplies to that country' (B5.40). In a confidential report for Dr Salazar dated 20 December 1965 – three days after the introduction of the UK Sanctions Order – Jardim stated that Portugal should 'adopt an attitude which could contribute positively towards assisting the Salisbury Government to face the consequences of the boycott' (B5.40). He added that 'if any goods reached the Mozambican ports bound for Rhodesia, including oil, service and transport would be made available'.[25]

Once the rail traffic had begun, the British Embassy officials who had mounted the watch at the Beit Bridge border were sent to monitor the flow of oil from Mozambique. On 19 February 1966 the Portuguese Foreign Ministry cabled the authorities in Lourenço Marques to warn them that two British diplomats, W. H. Harper and N. W. Lomas, had just applied for entry visas for Mozambique, and to advise that 'a discreet watch be kept on them'.[25] Two weeks later Foreign Secretary Michael Stewart summoned the head of the Portuguese Embassy in London to complain about this illicit trade. But Portugal stubbornly reaffirmed its hostility to sanctions, and the envoy announced that his country would continue to allow traditional rights of transit to its neighbours.

On 10 March 1966 the *Rand Daily Mail* reported that Rhodesia was receiving as much as 4,600 barrels a day of oil products by rail from Mozambique, a quantity which represented over half of Rhodesia's normal oil requirements, and which, with rationing, would have provided the bulk of the country's fuel needs. On the same day, Louis Walker informed Shell Centre in London that the rail traffic was increasing: 'Considerable quantities of products have been despatched by the Sonarep refinery destination Rhodesia. Our estimate is that some 80 to 90 rail tank-cars have been despatched over the past ten days' (B6.2).

*   *   *   *

The joke in Salisbury was that sanctions were now indeed 'biting' – not in Rhodesia, however, but in Zambia. The situation, as the US Ambassador in Lusaka explained, was 'as if Kaunda and Smith were to see who could hold their breath the longer'. It was, he added, 'a rather one-sided contest'.[27] Before the embargo, seventeen railway tank-cars had rumbled across the graceful bridge at Victoria Falls every day carrying essential oil supplies for Zambia. But when crude oil was cut off to the Umtali refinery, Rhodesia immediately halted all shipments of refined products to its northern neighbour. As Smith explained: 'If the British block up the entrance to a pipeline they should not express surprise when nothing comes out of the other end.'[28]

On 18 December 1965, the day after the Sanctions Order had been introduced, Zambia's supplies were cut off, and petrol rationing was immediately introduced. During the previous two months Zambia's stocks had been run down as the Rhodesian stockpile had been built up, and by the end of December the Copperbelt was on the verge of grinding to a halt. So low were stocks that at one point in early January 1966, Zambia was actually down to its last three days' supply of oil.

If Zambia was to obtain its oil requirements then there was no choice but to mount a massive airlift of oil supplies until sufficient stocks could be hauled in overland. The British Government agreed to provide this assistance. Just two days after the Sanctions Order was introduced, RAF Britannia aircraft, carrying 85 barrels of oil, began ferrying in supplies to Lusaka, The build-up was painfully slow. Airport facilities were inadequate, and soon began to deteriorate under the unaccustomed traffic. After the arrival of the last flight each evening, workers spread out over the runways to repair cracks that had opened up under the wheels of the heavily laden planes.

The introduction of the oil embargo had been delayed for over a month after UDI partly because plans needed to be made for the Zambian airlift. Wilson had sought assistance from the United States and Canada during his December visit, and both countries provided aircraft to help ferry in the oil. Within a few months sufficient stocks were built up inside Zambia. The US and Canadian airlift was ended on 30 April 1966, and the British effort continued for just a few weeks longer. The whole operation was extremely expensive, since more fuel was actually consumed by the planes than was transported by them, but the planes were refuelled at the coast so as to avoid a drain on Zambia's scarce stocks. When the RAF airlift was ended, in June 1966, nearly 200,000 barrels of oil had been flown in by the Britannias at a cost of £6 million to the British taxpayer.

Right from the introduction of the Sanctions Order road tankers were also used to haul oil more than a thousand miles along the Great North Road from the Tanzanian port of Dar es Salaam. The grandly titled road was actually nothing more than an unpaved dirt route for much of the way, and quickly deteriorated under the constant pounding of the heavy lorries, the verge soon becoming littered with broken-down wrecks. The lorry drivers themselves called it 'Hell Run'. Zambia, it seemed, was suffering even more than Rhodesia from the oil embargo.

The long-term solution to Zambia's problems was a pipeline from Dar es Salaam. Soon after the introduction of the oil embargo President Kaunda had written to Wilson suggesting that a pipeline could be built by the British army or by a UK company. Wilson's response had been cool. He replied that the pipeline would probably cost over £35 million, and would take a 'very long time' to complete, adding that construction of the Beira-Umtali pipeline, which was less than one-fifth of the length, had taken two years. 'I am sorry for the disappointing nature of this reply', the British Prime Minister concluded, 'but this is one of those matters where it is surely of the greatest importance to take account of the facts, however stark and unpleasant they may be.[29]

On the oil sanctions issue Britain seemed unable to put a foot right. Wilson's pessimism about the cost and construction of the pipeline proved quite unfounded. Eventually Zambia gave the contract to the Italian state oil company ENI which built the pipeline in seventeen months at a cost of £16 million – although after the Italians had been awarded the contract, there was an outcry that British firms had been discriminated against. When the Dar es Salaam-Ndola pipeline was opened in August 1968, Zambia's oil transport problems were solved.

*   *   *   *

The failure of the oil embargo against Rhodesia quickly became apparent as the weeks turned into months. While Zambia was still dependent on airlifted supplies, Rhodesia was importing its requirements by road from South Africa and by rail from Mozambique, with the result that on 3 March 1966 the Rhodesian Government was actually able to announce a partial relaxation of petrol rationing.

In spite of this, the British Government continued to give the impression that Smith's end was near. It seemed that the 'weeks not months' syndrome would never die. On 9 March 1966 the *Guardian* reported the views of the UK authorities:

> The British Government is confident that it is being supplied with full and accurate data on the kinds and quantities of petroleum products getting through to Rhodesia, and that these amount to the merest trickle. It is recognised in Whitehall that this does not square with accounts coming in from Salisbury and Johannesburg, suggesting there are large movements of oil by road and rail tankers, but the official view in London is that the Smith regime is going to some trouble to put on a good show for propaganda purposes. The world is being deliberately led to think that far more oil is crossing the Rhodesian frontier than the facts warrant.

The first few weeks of the oil embargo were absolutely crucial. It was then that Britain had the opportunity to try to nip sanctions-busting in the bud – and to show that it really meant business over the oil embargo. But it quickly

1

2

OILGATE –
THE UNCOVERERS

**The many faces behind the breaking
of the sanctions story:**

1 Jorge Jardim, Portuguese
  businessman and political advisor to
  Dr Salazar

2 Bernard Rivers and Martin Bailey
  (The Times)

3 Tiny Rowland, Chief Executive of
  Lonrho Ltd. *(Universal)*

4 Thomas Bingham Q.C. *(Universal)*

5 President Kenneth Kaunda of
  Zambia *(Popperfoto)*

3

4

5

MR. R.B. DUMMETT   11th April 1968

RHODESIA

The Portuguese Government have submitted a report to the French Government giving details of deliveries of oil from Lourenco Marques to Rhodesia. In a conversation some 10 days ago Dalemont told de Bruyne he would give him, in confidence, details of the report.

Duroc Danner, of C.F.P., has now given de Bruyne an opportunity to read the report which, I understand, C.F.P. intend to make available within the next two weeks to Fearnley of the Foreign Office.

The report gives details of R.T.C.s leaving Lourenco Marques with oil products for Rhodesia and shows that approximately 11,000 R.T.C.s were despatched in 1967 against approximately 5,500 in 1966. The detailed breakdown is as follows:-

	1966	1967
Shell	105	4,443
Caltex	288	1,697
Mobil	503	2,000
C.F.P.	869	1,256
SONAREP	664	1,400
S.A. Railways	3,000	240

Subject to Mr. McFadzean's agreement when he returns next Tuesday de Bruyne has agreed that the information should be made available as soon as possible to Alan Gregory of the Ministry of Power. Mr. Fraser asked me to keep you informed about this matter since he was most anxious that we should see the Ministry of Power as soon as the details became available in view of the undertakings given by the Managing Directors to Mr. George Thomson at their last meeting on 21st February.

A. H. SANDFORD

A letter from the Sandford File showing that Shell Mozambique
(jointly owned by Shell and BP) had supplied nearly half of
Rhodesia's oil in 1967.

GENTA
(PVT.) LIMITED

IMPORTERS AND EXPORTERS

FIRST FLOOR LINQUENDA HOUSE
SALISBURY
TELEPHONES 25113-4-7-8-0
TELEGRAMS 'GENTA'
P.O. BOX 8442   CAUSEWAY
RHODESIA
TELEX RH2228

8.4/12

m Danby 6/12   **SECRET**

m mackenzie

Our Ref : DA/EY/8135

3rd December, 1973.

Mobil Oil Rhodesia (Pvt) Ltd.,
P. O. Box 791,
SALISBURY

For attention : Mr. J. B. Nicol

Dear Sir,

      Referring to our letter 7852 of 31st July, 1973, we
attach a statement detailing estimated requirements of Petroleum
Fuels for the period January/April, 1974. These again are expressed
in cubic metres.

      Mr. Atmore has taken a copy of this statement with him
on his present visit to Capetown, but we shall be obliged if you
will also pass this information to your associates. The same
details have been supplied to our Agents in Lourenco Marques.

Yours faithfully,

D. Airey
OPERATIONS MANAGER

Encl.

A secret letter from Genta, the Rhodesian regime's oil
importing agency, to Mobil. This refers to the quantities of oil to
be sold to Rhodesia in 1974.

RHODESIA AND FREIGHT SERVICES

General

BP and Shell continue to market products in Rhodesia as a Consolidated venture. Supplies to support the marketing activity are
affected from South Africa primarily through "Freight Services",
who act as forwarding agents, buying product from BP and Shell S.A.
and reselling to the Rhodesia Government Procurement Agency GENTA
for allocation to marketers.

Consolidated sales to Freight Services in 1973 were as follows
(000 cbm).

Major Products	Mogas	174	
	Gasoil	189	
	TVO/Kero	40	
			403
Minor Products	Aviation	31	
	Lubes/Bitumen, etc		
	LPG	21	
			52
Total			455

Sales of major products in 1973 represent about 60% of total market
(circa 520 000 cbm major products) as against the 1972 volume of
314 000 cbm and 50% supply share. This increase does not represent
a typical or continuing situation, since it arises out of a "special
deal" which gave Consolidated a temporary increase in outlet in
Rhodesia at the voluntary expense of other marketers, including
supplies (it is believed) for Rhodesian reserve storage, and in
recognition of costs incurred by Consolidated in 1973 for construction
of improved rail loading facilities at Lourenco Marques. The supply
share is expected to return to a normal 50/51% thereafter, reflecting
market share in Rhodesia - Shell 38%, BP 13%.

Supplies

Following sanctions after UDI no crude oil could be imported to
Rhodesia to run the Umtali refinery, which was consequently shut
down. Alternative supplies of finished product were arranged
from South Africa under edict from Verwoerd who instructed that
normal trade relations with Rhodesia should be maintained. Devious
supply arrangements have thus been made to visibly disassociate the
oil companies from any first hand and identifiable part in supply
operations.

To achieve this, Consolidated now sell the bulk of products destined
for Rhodesia to "Freight Services" (Shell's forwarding agents) who
handle matters thereon by reselling to the Rhodesian Fuels Procurement Agency, GENTA, who onsell to the marketers in Rhodesia for
distribution and marketing.

**The Rounce Memorandum (from the Sandford File) which
explains in detail exactly how BP and Shell were busting
sanctions in February 1974.**

emerged that the British Government was quite unwilling to face the fact that Rhodesia was managing to import its oil in refined form through South Africa and Mozambique. This blindness to reality was soon to be even more vividly – and expensively – demonstrated in the dramatic confrontation that was about to take place off Beira.

NOTES

1 Wilson, *The Labour Government 1964–70*, p. 245.
2 Richard Crossman, *Diaries of a Cabinet Minister*, vol i, p. 407.
3 Robert Good, *UDI*, p. 107.
4 The Southern Rhodesia (Petroleum) Order 1965, section 1(1)(c).
5 Hansard, Commons, 20 December 1965, col. 1693–4.
6 Wilson, *The Labour Government 1964–70*, p. 256.
7 Good, *UDI*, p. 121.
8 Joe Garner, *The Commonwealth Office 1925–68*, (Heinemann, 1978), p. 394.
9 Interview, 28 March 1979.
10 Interview, 10 April 1979.
11 Chapman Pincher, *Inside Story* (Sidgwick & Jackson), p. 16.
12 Richard Hall, *The High Price of Principles: Kaunda and the White South*, (Penguin, 1973), p. 148.
13 *New Statesman*, 15 September 1978.
14 Interview, 5 April 1977.
15 Anthony Lake, *The 'Tar Baby' Option: American Policy Towards Southern Africa*, (Columbia University Press), p. 88.
16 Good, *UDI*, p. 130.
17 Hansard, Commons, 27 January 1966, col. 395–6.
18 Hansard, Commons, 27 January 1966, col. 396.
19 Good, *UDI*, p. 127.
20 Kenneth Young, *Rhodesia and Independence*, p. 381.
21 BBC Television, 19 September 1978.
22 *Guardian*, 19 February 1966.
23 Jardim, *Sanctions Double-Cross*, p. 42.
24 Jardim, *Sanctions Double-Cross*, p. 43.
25 Jardim, *Sanctions Double-Cross*, p. 32.
26 *Guardian*, 6 September 1978.
27 Good, *UDI*, p. 115.
28 Kenneth Young, *Rhodesia and Independence*, p. 361.
29 Richard Hall, *The High Price of Principles*, (Penguin, 1973), p. 148.

8 Blockade At Beira

April 1966

During the first two months of 1966 there were no less than thirty-two reports of tankers heading for Beira loaded with crude oil for Rhodesia. British diplomats were quick to dub them 'phantom tankers' – their existence, it was suggested, being nothing more than an elaborate hoax created by the rebel regime to give the impression that sanctions were failing. At this time, at least, the international press was keeping a close watch on the oil embargo, so much so in fact that there often seemed to be more journalists than dockers on the quayside at Beira.

It later emerged that a complicated plot was being hatched to import crude oil into Rhodesia. Wilson himself described the methods used as being 'more appropriate to the more fanciful flights of 007 than the kind of world in which the rest of us are living'.[1] This was a truly multinational scheme, involving a shipment of Iranian crude oil sold by the Lebanese subsidiary of an American company to two Greek firms. The oil was then to be carried in a Greek-registered tanker, owned by a Panamanian company, and chartered to a South African firm. The exercise involved a brisk turnover in ships' names, flags, captains, and destinations, and was a dramatic illustration of the difficulties of disentangling the labyrinth of international business interests at work in the sanctions-busting chain (see map of *Joanna V*, Fig. 2).

On 21 February 1966 the *Arietta Venizelos* sailed from the Iranian port of Bandar Mashur bound for Rotterdam. At this point there was nothing to distinguish this particular tanker. It carried a cargo of 110,000 barrels of crude oil originally belonging to the San Jacinto, a Beirut-based subsidiary of US Continental Oil. Nicos Vardinoyannis, a

146

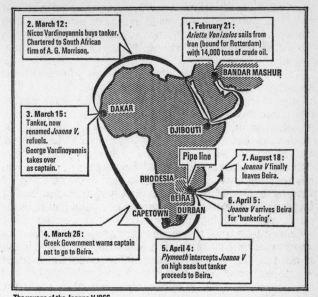

Fig. 2

The tortuous voyage of the *Joanna V* around Africa in 1966. This led to the establishment of the Beira Patrol, and marked Rhodesia's final attempt at importing crude oil. From then on Rhodesia purchased its oil requirements in refined form from Mozambique and South Africa.

Greek entrepreneur, was the purchaser of the oil, which had been acquired by two of his companies – Seka and Nima International. He had already arranged to charter a tanker from Venizelos, another Greek shipping firm, to carry the cargo.

On 8 March, while the tanker was steaming through the Mediterranean, Vardinoyannis attempted to purchase the vessel from the company that had chartered it to him. His offer of £400,000, about double its nominal value, was quickly accepted. 'As long as they're buying our ship for money like that,' Nikitos Venizelos explained, 'who asks questions?' On 12 March Vardinoyannis then arranged for the actual purchase to be made by one of his Panamanian subsidiaries, Varnikos Corporation, and the transaction was completed in New York with a letter of credit from Johannesburg.

Although the Beira-Umtali pipeline was still closed, the Rhodesians seemed confident that they would soon be able to import crude oil, the Minister of Commerce having announced on 25 February 1966 that a tanker would arrive at Beira 'in the foreseeable future'.[2] Soon afterwards he proclaimed that 'the day our first tanker arrives in Beira we shall have won this economic war.[3] This later proved something of an exaggeration: the Portuguese authorities had first to permit the oil to be unloaded; it then had to be pumped the short distance from the wharf to the terminal of the pipeline; and finally the pipeline company had to agree to transport the oil into Rhodesia.

Cooperation from the pipeline company was key. Smith therefore personally summoned Rowland, the chief Lonrho Director of the pipeline company, for an important meeting in Salisbury on 8 March. Rowland later recalled that 'the interview made rapid progress from the persuasive to the threatening'.[4] If the Lonrho chief did not authorise the construction of a new connection from the wharf to the Umtali pipeline, which would be necessary if the oil was to be pumped, then Rowland 'would be jailed' (B5.68). Rowland claims that at the end of 'this unpleasant meeting' he was removed under guard to Brigadier Dunlop's Ministry of Transport before being released.[5]

Frantic efforts were meanwhile being made at the port of Beira to receive the oil already on its way in the *Arietta Venizelos*. But one of the problems was that it was not normally possible to pump the oil directly from a tanker into the Umtali pipeline. Supplies therefore had to go through the storage tanks belonging to an international oil company before they could be despatched to Rhodesia. These companies had made it clear that they would not cooperate with any attempt to pump crude oil to Rhodesia. Urgent measures were therefore taken by the Rhodesian authorities to set up a company with the innocuous name of Mocambique Investmentos Limitada, to erect six prefabricated storage tanks at Beira. These tanks would have to be filled from a short pipeline owned by Mozambique Railways, which ran along the quayside used by oil tankers. The Mozambique authorities, however, had removed a small part of the connecting pipeline – a crucial T-piece – so that no transfer of crude oil could actually take place without their cooperation.

Meanwhile, on the high seas, the *Arietta Venizelos* received new instructions. Vardinoyannis, the new owner of the tanker, chartered his vessel to the Cape Town company of A. G. Morrison. Instead of sailing to Rotterdam, as originally planned, the tanker was ordered southwards to the Senegalese port of Dakar for refuelling. A few days later it sailed under a different captain – George Vardinoyannis, the younger brother of the new owner, who flew out from Athens to captain the tanker on its delicate mission to Southern Africa. The tanker, now renamed the *Joanna V*, was about to hit the headlines of the international press.

British intelligence sources had already become interested in the movements of the *Joanna V*. Because the vessel had been registered at Piraeus, the Greek Government was alerted, and the new captain was pursued by the radio badgerings of the Greek maritime authorities on his tortuous trip round Africa. George Vardinoyannis was specifically warned that it was illegal for a Greek vessel to deliver oil at Beira for Rhodesia. But by the end of March, the *Joanna V* had rounded the Cape, and was fast approaching the Mozambique Channel.

British naval units were now in hot pursuit. A Shackleton reconnaissance airplane, after spotting the *Joanna V*, immediately flashed the news to a patrolling frigate, the *Plymouth*, which intercepted the tanker on the high seas during the evening of 4 April. The *Joanna V*'s captain explained that he had been instructed to put into Beira for bunkering and provisions, and then to proceed up the coast to Djibouti to discharge her cargo. It seemed an unlikely story: Djibouti itself had no oil refinery, and in any case the tanker had passed within a few miles of the port nearly six weeks earlier before it had circumnavigated the whole African continent.

The Royal Navy had no power to prevent tankers sailing for Beira. This could only have been done with the prior approval of the government which had registered the vessel, and the Greek Foreign Ministry was only willing to give such authority if specifically requested by the United Nations. The captain of the *Plymouth* therefore had little option but to allow the tanker to proceed. At 5.20 in the morning of 5 April 1966 the *Joanna V* dropped anchor just over a mile offshore from the quay at Beira – a blatant symbol of defiance against Britain's Rhodesian policy.

*   *   *   *

On the day the *Joanna V* arrived at Beira, Foreign Secretary Michael Stewart summoned the head of the Portuguese Embassy in London to request Portugal to refuse to allow the tanker to discharge. Meanwhile Lord Walston, Parliamentary Under-Secretary at the Foreign Office, left on an urgent mission to Lisbon with a similar request for the Portuguese Foreign Minister. Although Portugal refused this request, Lord Walston was assured that the emergency storage tanks under construction at the port would not be used to handle oil for Rhodesia.

Portugal faced a painful dilemma. Dr Salazar was sympathetic towards the Smith regime, but Portugal could not afford to be cast in the role of villain in defying the United Nations. For if the Portuguese Government actively assisted Rhodesia, there was a danger that sanctions might be extended to cover Portugal itself. The open supply of crude oil to the Umtali refinery, Jardim later explained, would have given Wilson 'an excellent reason to blame Portugal before the world for the failure of sanctions'.[6]

'The United Kingdom has turned the arrival in Beira of a tanker into a symbol,' an internal Portuguese Government report concluded on 19 March 1966. 'Very deliberately, Britain has been diverting attention from the behaviour of the other countries helping Rhodesia to focus international attention on Beira . . . [Portugal's] aim is essentially to gain time by not giving any excuse for international intervention.'[7] Clearly, the Portuguese Government had to tread carefully.

The Ministry of Foreign Affairs in Lisbon issued a long statement on the affair on 6 April. 'The Portuguese Government,' the Ministry emphasised, 'is not prepared to assume the role of culprit in a situation for which it formally and

150

categorically rejects any responsibility.'[8] Portugal would not take any positive initiative to encourage the supply of oil to Rhodesia. But neither would it prevent the consignment of oil which was being sent by others in transit through Mozambique.

Wilson's position was becoming increasingly desperate. 'News had reached us,' he later explained, 'of a determined exercise by the [Rhodesian] regime, together with some members of the shipping underworld, to run the Beira gauntlet.'[9] Clearly, the *Joanna V* was merely blazing the trail for a long stream of tankers. One source suggested that the Cape Town company of A. G. Morrison had signed a contract for as many as twenty-seven cargoes of crude oil totalling 3,200,000 barrels, which would have been sufficient to last Rhodesia for a full year.[10]

Already a second tanker was fast approaching Beira. It too had sailed from the Iranian port of Bandar Mashur, on 27 March 1966, with a cargo of 110,000 barrels of crude oil. Shortly before its departure, the tanker had also been purchased by Vardinoyannis' Panamanian company, Varnima Corporation, and like his earlier acquisition, it sailed under the Greek flag. The vessel, previously named the *Charlton Venus*, was immediately christened the *Manuela* in honour of the new owner's wife (see map of *Manuela*, Fig. 3).

On 8 April the *Manuela* sailed past the approaches of Beira and into the political storm whipped up by the *Joanna V*. The captain stated that his destination was Rotterdam, via Durban. But the following day the *Manuela* suddenly turned, and headed north again towards Beira. The two Greek tankers were driving an enormous breach through the oil embargo; and no doubt where they led, other tankers would soon follow.

'Had we left matters where they were,' Wilson later explained, 'sanctions would have become totally ineffective and our Rhodesia policy would have been destroyed.'[11] The Prime Minister added that 'pressure for the use of force would have become stronger and stronger, with incalculable results, particularly if Eastern-bloc countries joined in the struggle on the ground'.

It was the most serious foreign crisis which Wilson had faced. The *Joanna V* had arrived at Beira, and might attempt to discharge its cargo at any time, while the *Manuela*

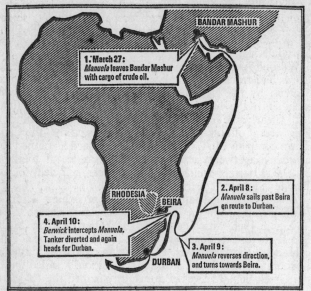

BANDAR MASHUR

1. March 27:
Manuela leaves Bandar Mashur with cargo of crude oil.

RHODESIA

BEIRA

2. April 8:
Manuela sails past Beira en route to Durban.

4. April 10:
Berwick intercepts *Manuela*. Tanker diverted and again heads for Durban.

3. April 9:
Manuela reverses direction, and turns towards Beira.

DURBAN

The voyage of the Manuela 1966

Fig. 3

The *Manuela* tried to deliver oil to Beira in the wake of the *Joanna V*. The tanker was diverted by the Royal Navy.

was still on course for the port and would be arriving within a matter of days. If Wilson did not pre-empt African demands for the use of force, by taking a tough stand over the oil embargo, then the British Government would soon be facing even greater pressure at the United Nations. Britain decided to act.

The Labour Government, having been returned to power on 31 March with an overall majority of 97, was now in a position to take tougher measures over Rhodesia. On the morning of 7 April 1966 the Cabinet met and decided to seek authorisation from the United Nations to stop the delivery of oil to Beira. Lord Caradon, Britain's Representative at the UN, was instructed to call an immediate meeting of the Security Council. In New York a majority of the Council's fifteen members were prepared to meet, some delegates having actually assembled in the chamber at 10 am. But the President for the month, the Representative of Mali, was nowhere to be found, and under the UN Charter the Security Council could not formally meet without its President.

Britain still had no authority to prevent the *Joanna V* unloading or the *Manuela* from entering Beira. The Government was therefore totally hamstrung by the delay in convening the Security Council. 'Throughout Good Friday and the Saturday,' Wilson later explained, 'I was receiving ominous telegrams over the Number Ten secret teleprinter to my holiday home.'[12] As the precious hours slipped by, Lord Caradon became increasingly impatient. Mali appeared to be delaying the UN meeting out of anger that Britain had originally refused to use force to quell the Rhodesian rebellion immediately after UDI. It was not until two days later, on Easter Saturday, that the Mali Representative appeared in the chamber, and the Security Council finally assembled to consider action against the pirate tankers.

Lord Caradon opened the debate by stressing the urgency of the crisis. 'As we meet here today,' he explained, 'the *Joanna V*, with a full cargo of oil, rides at anchor in the port of Beira.'[13] In addition Royal Navy aircraft had just reported that the *Manuela* had altered direction and was now 'on a course which could take her to Beira in about twenty-four hours'. The British Representative therefore submitted a draft resolution which would authorise the United Kingdom to use military force to prevent the arrival of oil tankers at Beira.

The African members of the Security Council pressed for tougher sanctions. Why, they asked, had Britain not used force to crush the rebellion at the time of UDI? And why were British efforts concentrated solely on the supply of oil through Beira? If the delivery of oil to Beira represented 'a threat to the peace,' it was argued, then surely this should also apply to oil supplied from Beit Bridge and Lourenço Marques?

The African members of the Security Council therefore tabled a number of amendments to the British draft. One of these called upon South Africa 'to take all measures necessary to prevent the supply of oil to Southern Rhodesia'. This additional clause, the Ugandan Representative noted, did not actually state that South Africa was breaking the oil embargo. 'If it has been doing so, then the cap fits; if not, all we are asking South Africa is that it should take all measures to prevent the supply of oil to Southern Rhodesia. This is not too much to ask.'[14]

Voting took place at 9 pm on Easter Saturday. Lord Caradon announced that he was unable to accept any of the African amendments without 'further instructions' from London and that since the matter was urgent, he would be forced to abstain. The other Western members of the Security Council also abstained, and none of the African amendments obtained sufficient votes for approval. The original British draft resolution was passed by ten votes to none, with five abstentions. Mali, the Soviet Union, and Bulgaria abstained on the grounds that it did not go far enough towards dealing with the Rhodesian crisis; France claimed that Rhodesia was not a true threat to international peace and, in any case, was an internal affair of Britain, and therefore not within the competence of the United Nations; the representative of Uruguay abstained because he had not received instructions from his government.

The Beira Resolution called upon the Portuguese Government not to allow oil for Rhodesia to be landed at Beira or to permit oil to be pumped through the pipeline to Umtali. All states were instructed to divert 'any of their vessels reasonably believed to be carrying oil destined for Rhodesia'.[15] Most importantly, the resolution called upon Britain 'to prevent by the use of force if necessary the arrival at Beira of vessels reasonably believed to be carrying oil destined for Rhodesia, and empowers the United Kingdom to arrest and detain the tanker known as the *Joanna V* upon her departure from Beira in the event her oil cargo is discharged there'. Nicos Vardinoyannis, the Greek owner of the two rogue tankers, was furious. 'This is like a gambler who loses at poker with a hand of four aces because his opponents has two guns,' he declared.[16]

The Beira Resolution marked an important turning point in UN history. It was the first occasion since the Korean War when a dispute had been declared a 'threat to international peace', giving the Security Council, under Chapter VII of the UN Charter, the right to authorise the use of force. The Beira Resolution, after expressing grave concern at reports that crude oil supplies might be delivered for Rhodesia, determined that 'the resulting situation constituted a threat to the peace'. The acceptance of this phrase marked a distinct shift in British policy.

* * * *

The results of the Security Council vote were immediately flashed to the Royal Navy. Early in the morning of 10 April 1966 the frigate *Berwick* intercepted the *Manuela* in the Mozambique Channel. The frigate signalled to the tanker, ordering her not to proceed to Beira, but the *Manuela* continued to maintain her northerly course. The *Berwick* then hoisted an international signal to announce that she was despatching a boarding party. Two officers and an armed escort of two boarded the tanker. Even then, however, the *Manuela's* captain refused to alter course. Only after a back-up party of twelve armed men was put aboard did the tanker turn around and proceed on a southerly course towards Durban.

Meanwhile, back in Beira, the *Joanna V* moved a little closer to the pipeline. On 11 April the tanker pulled anchor and came in to dock at Quay Number Eight. The quay was only ten yards from the pipeline, and rumours quickly spread that an attempt was being made to obtain a 'reduction valve' that would permit a direct connection between the *Joanna V*'s six-inch off-loading pipe and the ten-inch pipeline.

But the *Joanna V* was already an international outcast. As soon as the tanker arrived at Quay Number Eight, the Greek Consul in the port elbowed his way through a crowd of exuberant Rhodesian holiday-makers to board the vessel. He informed the captain that the Greek authorities had withdrawn the tanker's registration five days earlier and that the *Joanna V* was now a pirate ship. A few hours later a crew member appeared over the side of the tanker to paint out the word 'Piraeus' – the old port of registry – and replace it in white paint with the word 'Panama'. The following day, however, the Panamanian Government withdrew the tanker's provisional registration. The same painter lowered himself over the stern, this time to black out the word 'Panama'.

The spotlight of international attention was now on Portugal. 'Our position was not at all comfortable,' Jardim recalled, and Dr Salazar 'thought that the problem should be put to Mr Ian Smith to make the final decision.'[17] Jardim was therefore sent to Salisbury for discussions with Smith. The meetings were lengthy, lasting for over four hours, and Jardim carefully explained the Portuguese position: Portugal wanted to uphold the rights of free transit. But to allow the *Joanna V* to unload its controversial cargo would simply lead to an escalation of the crisis. Besides, Jardim added, Rhodesia could continue to import refined oil products through Lourenço Marques.

On 16 April 1966 Smith announced his decision not to use the oil from the *Joanna V*. This decision, he explained, had been taken in order to avoid drawing Portugal into the dispute between Britain and its rebel colony. Smith added that Rhodesia had only begun importing oil through the pipeline the previous year. 'Prior to that we got along very well using other means' – a reference to the importation of refined products by rail – 'and so we will make do and continue by using these other traditional lines of supply' (B5.73).

It was not until 18 August 1966, more than four months later, that the *Joanna V* actually left Beira. The tanker could not have safely sailed fully-laden because a sandbank blocked the entrance to the port, and the British Government therefore gave permission for 16,000 barrels of the crude oil to be pumped into another tanker. The *Joanna V*, Wilson later explained, was then 'courteously escorted by the Royal Navy away from all the places where her presence was not desired'.[18] The pirate tanker had finally sailed off the pages of the world press.

*　　*　　*　　*

After the Beira Resolution, the Royal Navy began the long task of patrolling the Mozambique Channel. The Admiralty arranged for ships to stop off in the Mozambique Channel for a spell of patrolling on their way to or from a tour of duty in the Far East. Shackleton reconnaissance planes, based in

Madagascar, were also used to check the sea lanes for a number of years, and a Tanker Unit was set up in London in 1966 to compile a list of 'innocent' tankers which was then passed on to the patrolling frigates. Other tankers were liable to be intercepted if they made for Beira. There were ten interceptions during the rest of 1966, and fourteen in the following year.

In December 1967 a Royal Navy frigate actually fired four shots across the bows of a French oil tanker, which turned out not to be carrying crude oil to Beira. Gradually, however, the number of interceptions declined, and no tankers were stopped after 1972. The number of patrolling vessels also steadily decreased, and from March 1973 the patrol was only operated on an intermittent basis. By then the risk of a tanker trying to make for Beira was small, since the Umtali refinery could only have resumed operations after a lengthy recommissioning period.

For the sailors, the Beira Patrol was an unpopular assignment. One naval officer later described how the men would settle down to 'a fortnight of intense boredom, relieved only by kite-flying competitions, the dropping of mail into the sea from ancient Shackletons, while watching with increasing apathy as the ships passed unmolested'.[19] He added that the only benefit of the tour of duty off Beira was 'the political education of the participants'. As part of their 'political education', the sailors saw at first hand how pointless and wasteful it was to patrol Beira, while the oil was being shipped in through Lourenço Marques. Smith too must have laughed at the charade.

When the Bingham Report was published in 1978 there was a flurry of angry letters to the press from retired naval officers. Lieutenant Commander Gidley Wheeler, for instance, wrote to *The Times* 'to give vent to the white hot anger' that he felt on being 'misused and deceived by politicians and civilians'.[20] He ended his letter on an ominous note, warning that when the politicians abuse soldiers, 'they tread a dangerous path . . . ' Undoubtedly the Beira Patrol sapped morale in the forces. Indeed on one well-publicised occasion one of the blockaders left to join the rebels: Mike Mason left the Royal Navy in 1972 to become a civil servant for the Smith regime in Rhodesia.

Portugal has traditionally been Britain's oldest military

ally, a fact which merely added to the anger of the Royal Navy sailors. But there was an unusual degree of goodwill between the navies of the two nations which were stationed in the Mozambique Channel, as was illustrated by an incident that occurred during the football World Cup when Portugal and England met in the semi-finals. Jardim has recounted how sailors on the Portuguese and British naval vessels had been glued to their radios. 'Excitement rose, but calm was restored when the captain of the frigate *Alvares Cabral* congratulated the skipper of the nearest British vessel on the English victory. An immediate reply came by semaphore acknowledging the greetings and praising the Portuguese for having played an excellent match.'[21]

The last British vessel to patrol off Beira was the aptly-named *Salisbury*. On 25 June 1975, the day of Mozambique's independence, the patrol was formally ended, Britain arguing that the newly-independent state had undertaken to ensure that crude oil supplies to Rhodesia were not resumed without authorisation from the UN Security Council. By 1975 fifty-two tankers had been stopped, although none of them had been en route for Beira with crude oil. A total of seventy-six ships and 24,000 men had at one time been involved in the ten-year exercise. The total costs of the patrol are difficult to estimate, but they probably exceeded £100 million.

'Staggering sums were squandered,' Jardim pointed out, 'to keep one of the biggest bluffs in contemporary history on stage.'[22] The patrol had acted as a powerful deterrent against any attempts to send crude oil to Rhodesia. But it did absolutely nothing to halt the flow of refined oil products. Dr Owen, as Opposition spokesman on defence, pointed out that 'it is extraordinary that it was thought necessary to achieve this limited objective, of preventing crude oil reaching Rhodesia, in such a flamboyant manner, when a similar result could have been achieved by covert action, such as blowing up the pipeline or sabotaging the pumps'.[23] Covert action by British agents on Mozambican territory would have had extremely serious diplomatic implications. But even so it is perhaps surprising that Britain should not have seriously considered limited 'force' inside Rhodesia to ensure that the Umtali refinery never resumed operations.

The Beira Patrol came under considerable attack in Parliament, particularly from Conservative MPs who were hostile towards sanctions. But successive Governments – both Labour and Conservative – kept up the patrol until the independence of Mozambique for fear that a withdrawal of the Royal Navy would have focused further international attention on the failure of the oil embargo. There were other fears in the mind of Government: in Salisbury the ending of the patrol would have been regarded as a gesture of surrender to the Smith regime; at the United Nations it would no doubt have led to strong pressure for further measures to tighten sanctions; and to withdraw the Navy might also have offered the Soviet Union an opportunity to move its naval units into the Mozambique Channel. The Beira Patrol, once it had begun, would have been difficult to end as long as Portugal ruled Mozambique. For a decade the naval charade continued.

*   *   *   *

The UN had called on Britain to 'prevent by the use of force if necessary the arrival at Beira of vessels reasonably believed to be carrying oil destined for Rhodesia'. The resolution therefore applied to both *crude* and *refined* oil. In the case of crude oil, any supplies delivered to Beira could only have been destined for Rhodesia, since there was no refinery in the Mozambican port. But the position with regard to refined oil products was more complicated, since these were also consumed in Mozambique. Nevertheless, very little effort seems to have been made to determine which consignments of refined oil products were being sent on to Rhodesia.

Refined oil products destined for Rhodesia actually continued to reach Beira under the very eyes of the Royal Navy. Employees of Shell Mozambique have claimed that the company continued to send specialised oil products (such as lubricants and paraffin) into Rhodesia from Beira. In February 1974 a BP report confirmed that both Shell and BP were supplying 'minor quantities' of bitumen through Beira; and a Mobil memorandum dated 14 June 1971 suggested that lubricants, greases, and paraffin waxes were being sent to Mobil Rhodesia through the same route (B8.39).[24]

159

The British Government clearly knew about the shipment of refined oil products through Beira. In March 1968, for instance, the Foreign Office received a telegram reporting that a South African ship at Durban had loaded one hundred Total oil drums which were being sent through Beira to Salisbury by the British-owned forwarding firm of Mitchell Cotts (B9.8). Britain also had a Consul at Beira who was responsible for reporting on sanctions-busting. It is not known how much information John Taylor, and later René Howell, sent back to London. But one employee of Shell Mozambique has recalled that when he played tennis with the British Consul at the Country Club in Beira, 'the question of oil supplies was always diplomatically avoided'.

Bingham confirmed that 'it seems probable that there was some flow of refined products to Rhodesia through Beira' (B9.12), but considered that in the case of Shell and BP, at least these did not make 'any substantial contribution towards satisfaction of Rhodesia's need for these products. Nevertheless, the fact that all the international oil companies continued to send some refined products to Rhodesia through Beira, with no action being taken by the British authorities, was yet another sign of the weakness of the British Government's attempt to enforce the embargo.

'The Beira Patrol,' Dr Owen wrote, 'was a classic illustration of the failure of politicians to look realistically at the facts and of their tendency to rely on optimistic projections of that which they wish to hear.'[25] The British Government tackled the problem of cutting off crude oil supplies energetically, and the setting up of the Beira Patrol was a decisive attempt to make the embargo effective. But the British Government seemed to be blind towards the increasing flow of refined oil through the port of Lourenço Marques some six hundred miles to the south. Britain had won the battle of Beira, but it was rapidly losing the oil sanctions war.

NOTES

1 Hansard, Commons, 21 April 1966, col. 100.
2 *Sunday Times*, 10 April 1966.
3 Good, *UDI*, pp. 132 and 135.

4 Letter from Rowland to Edmund Dell (Department of Trade), 7 January 1977, p. 17.
5 Letter from Rowland to Edmund Dell, 7 January 1977, p. 17.
6 Jardim, *Sanctions Double-Cross*, p. 67.
7 Jardim, *Sanctions Double-Cross*, pp. 69–70.
8 Jardim, *Sanctions Double-Cross,* p. 72.
9 Wilson, *The Labour Government 1964–70*, p. 289.
10 *Sunday Times*, 10 April 1966.
11 Wilson, *The Labour Government 1964–70*, p. 290.
12 Wilson, *The Labour Government 1964–70*, p. 291.
13 Security Council debate, 9 April 1966.
14 Security Council debate, 9 April 1966.
15 Security Council resolution 221 (1966).
16 *Sunday Times*, 10 April 1966.
17 Jardim, *Sanctions Double-Cross*, p. 77.
18 Wilson, *The Labour Government 1964–70*, p. 291.
19 *Sunday Times*, 3 September 1978.
20 *The Times*, 26 September 1978.
21 Jardim, *Sanctions Double-Cross*, p. 84.
22 Jardim, *Sanctions Double-Cross*, p. 84.
23 Dr David Owen, *The Politics of Defence*, (Cape, 1972), pp. 114–5.
24 *The Oil Conspiracy*, pp. 45–8.
25 Dr David Owen, *The Politics of Defence*, (Cape, 1972), p. 114.

9   An Embarrassing Discovery

May 1966–January 1968

One of the most curious aspects of the sanctions story is that it apparently took the British Government so long to discover that Shell and BP were actually supplying half of Rhodesia's oil. Even more incredible, however, is the claim of the head offices of Shell and BP in London that they did not learn that their Southern African subsidiaries were busting sanctions until over two years after UDI. Either the Government and the oil companies were astonishingly ignorant – or the facts were known and suppressed.

Halting the supply of crude oil, Bingham concluded, was 'the first, perhaps the only, major victory won by the oil sanctions policy' (B5.75). But while the British Government was so energetically preventing the *Joanna V* from delivering *crude* oil to Beira in April 1966, massive supplies of *refined* oil products were being sent into Rhodesia by other routes. Initially the British Government gave the impression that the flow of refined oil was minimal. Just three days after the Security Council had approved the Beira Resolution, a Foreign Office spokesman was quoted as saying that the logistical problems of sending in refined products to Rhodesia would be 'tremendous', and would make the operation 'frightfully expensive'.[1]

The battle against the *Joanna V* had hardly satisfied the militant African members of the UN. Why concentrate solely on preventing crude oil reaching Beira, they asked, while you virtually ignore the flow of refined oil products from South Africa and Mozambique? As the Representative of Pakistan explained, it was like 'stamping a mouse in the kitchen whilst a lion was roaring at the door.'[2] Nigeria therefore proposed a Security Council resolution calling on the United Kingdom 'to prevent any supplies, including oil and petroleum products from reaching Southern Rhodesia.'[3] Britain and the other Western members of the Security Council abstained on the vote, and the Nigerian proposal was defeated, having failed to win sufficient support.

Lord Caradon, the British Representative, had rejected the Nigerian resolution on the grounds that sanctions were already beginning to bite: 'The regime is now very concerned at the cost of maintaining its oil supplies,' he explained, 'and the oil embargo is adding very seriously to its financial difficulties.'[4] Those who believed that the oil embargo had already failed, he claimed, were victims of Smith's propaganda machine:

> No wonder that the regime does not dare tell its own people the facts . . . All the more salutory and sudden will be the effect when this curtain of concealment is torn aside and when people in Rhodesia face the inevitable truth. The hollowness behind the campaign of secrecy to which they have been subjected will then be fully apparent.[5]

As it turned out, however, it was not the Rhodesian regime which was concealing the truth about sanctions but Britain itself.

*   *   *   *

By the time of the introduction of the Beira Patrol most of Rhodesia's oil was being sent by rail along the South African Loop Route. Railway tank-cars were loaded with refined oil at the port of Lourenço Marques, and then despatched into South Africa. Shortly after they had crossed the South African border at Komatipoort, the tank-cars were suddenly turned back the way they had come and returned to Mozambique. From there they were consigned to Rhodesia as if they had originated from South Africa (see map of South African Loop, Fig. 4a). This complicated route made it more difficult to uncover exactly who was involved in supplying Rhodesia's oil.

During 1966 a total of 700,000 barrels of oil was despatched from South Africa to Rhodesia by rail. At the time it was not generally suspected that the oil companies themselves were involved in this trade; but Bingham revealed that

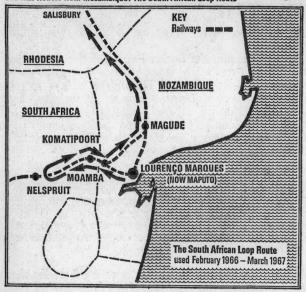

The South African Loop Route
used February 1966 – March 1967

around *two-thirds* of this oil was actually sent by the South African subsidiaries of Shell and BP (B7.5). From the start a thinly disguised paper-chase was used: Shell and BP would sell a consignment of oil to the South African firm of Parry Leon & Hayhoe (later absorbed into Freight Services), who would then arrange onward shipment to Rhodesia.

The involvement of Shell and BP at this very early stage puts the whole sanctions story in a fascinating perspective. It shows that only a few months after the introduction of the oil embargo, the oil companies had already become the major sanctions-busters.

Arthur Bottomley, who was Commonwealth Secretary at the time, has claimed that he was never told that the British-owned companies actually supplied two-thirds of the oil sent in on the South African Loop. 'If I had known,' he added, 'then I would – in the strongest terms – have demanded an explanation from Shell and BP.'[6] It is fascinating to specu-late how very different the whole course of the Rhodesian

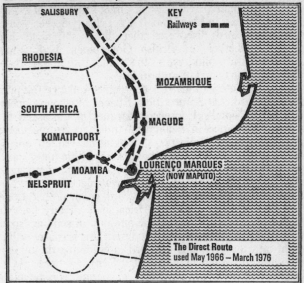

SALISBURY
KEY
Railways ▬▬▬

RHODESIA

MOZAMBIQUE

SOUTH AFRICA

MAGUDE

KOMATIPOORT

LOURENÇO MARQUES
(NOW MAPUTO)

MOAMBA

NELSPRUIT

The Direct Route
used May 1966 – March 1976

rebellion might have been if the Government had discovered the crucial fact of the involvement of British-owned companies back in 1966.

The South African Loop, although considerably cheaper than road transport, still proved expensive, since the greater mileage involved additional transport costs. Customs duty was also charged on the oil when it briefly crossed into South Africa, and the shipments took longer to reach their destination. Genta, the Rhodesian procurement agency, therefore suggested that to avoid the tortuous journey via South Africa the oil should be railed directly from Lourenço Marques into Rhodesia (see map of Direct Route, Fig. 4b).

In May 1966 George Atmore, Genta's Chairman, approached Jardim with the news that 'the international companies were willing to use this route,'[7] and asked whether Sonarep would make the first shipment, 'just to check a possible British reaction'; apparently Henry Downing, the British Consul at Lourenço Marques, 'had got

165

informants at the railway and at the port' (B7.18). Jardim agreed, and Sonarep sent the first direct shipment of twelve railway tank-cars in May 1966. It is not clear whether Downing's 'informants' discovered that direct shipments had begun, or whether the British Government had simply decided to turn a blind eye to the traffic. Either way, there was no reaction from UK authorities.

The other international oil companies quickly followed Sonarep's lead. On 3 June 1966 Eugenio Lisboa, General Manager of Total's Lourenço Marques office, asked for Jardim's assistance in facilitating transport arrangements with Mozambique Railways, scribbling his request on the back of one of his tiny visiting cards. This little note helped open the route which the international oil companies were to use to supply Rhodesia over the following decade.

The note on Lisboa's visiting card recalled that Louis Deny, head of Total South Africa, had already been in touch with Jardim to discuss the question of sending 'some gallons' of petrol directly to Rhodesia. This was a revealing insight into the problems involved in enforcing the oil embargo, for only a few months earlier Deny had been told by his superior at head office *not* to permit shipments to Rhodesia (see p. 138). Either the French oil executive was disobeying his orders, or else Total's Paris headquarters had issued new instructions allowing trade with Rhodesia. In any case, the upshot was that Total's South African subsidiary despatched its first consignment of fifteen railway tank-cars for Mozambique to Rhodesia later in May.

Later Mobil became the first American-owned company to use the Direct Route. But initially Fernando Gomes, head of Mobil's Lourenço Marques office, had been worried that Mozambique customs documents might implicate his firm in this sanctions-busting traffic and had approached the authorities to suggest changing the customs procedure in order to disguise Mobil's trade with Rhodesia. The plan was outlined in a secret cable sent by the Governor General of Mozambique to Lisbon on 9 August 1966. It explained: 'Mobil seeks to pass possession of Rhodesia-bound fuels to local transit firm Parry Leon, which in turn will forward them to their destination' in unmarked Rhodesia Railways tank-cars (B7.41). This would have made it more difficult to prove that Mobil was busting sanctions.

166

Mobil's request was rejected. The Overseas Minister told the Mozambique Governor General that when Mobil delivered products to a transit firm, 'the final destination should be declared immediately, so that the company will not be able to declare ignorance on this subject afterwards' (B7.41). In this way the Portuguese Government could show that it was the oil companies – not Portugal – which were fuelling Rhodesia. On 20 August the Mozambique Governor General confirmed to Lisbon that the transit agent (Parry Leon & Hayhoe) would have to declare the final destination of the oil on the customs clearance form, and that Mobil could only deliver oil after receiving a copy and registering it in its storage book. Mobil nevertheless decided to proceed, and a few days later the company despatched its first consignment of ten tank-cars directly to Rhodesia.

Caltex was the next company to join the Direct Route. In June 1966 Caltex South Africa had apparently offered to supply Rhodesia with oil at a discount – on one condition: that the company acquired a monopoly as Rhodesia's sole supplier (B5.92). Although this offer was later withdrawn, much to the anger of the Rhodesians, Caltex remained anxious to supply the Rhodesian market by the Direct Route. Caltex, however, faced an additional logistical problem, since its South African refinery was located at Cape Town, nearly 1,300 miles away from Lourenço Marques. A swap arrangement was therefore made with the Shell/BP and Mobil at Durban, under which Shell/BP and Mobil refineries would supply oil products to Caltex in Lourenço Marques for onward railing to Rhodesia, and in return Caltex would provide the other companies with matching quantities of oil to market in the Cape Town area. In this way everyone saved on transport costs.

Shell and BP had a delicate legal problem which was not shared by their American and French rivals. The operations of the other international oil companies – Total, Mobil, and Caltex – were conducted in Mozambique by their South African subsidiaries, which were registered in Cape Town, and therefore not covered by American or French sanctions legislation. But for historical reasons, the jointly-owned Shell/BP subsidiary, Shell Mozambique, was registered in Britain, and consequently subject to the UK Sanctions Order. In addition, most of the Directors of Shell Mozambique were British citizens, and they too were under a legal obligation to respect the embargo. This meant that Shell Mozambique was contravening UK law by making direct shipments to Rhodesia.

Eventually Shell and BP followed the other oil companies, starting direct shipment on 8 December 1966. But the fact that they hesitated before using this route was not due simply to the legal complications; for Shell and BP supplied around two-thirds of the oil sent on the South African Loop, and as Bingham has commented, the two British-owned companies may well have been the last 'to join the direct traffic because of their heavy involvement in the indirect trade' using the other route (B7.62). Soon, however, Shell and BP recognised the advantages of the Direct Route, and saw that they would lose business if they continued to send their oil on the Loop. As Jacobus Louw, the Shell South Africa Director responsible for sales, had argued: 'It was high time that we caught up with the others and we can't see our share of the market going down the drain (B7.29).

By the end of 1966 the basic supply procedures which were to be used for the following decade had been established. The local subsidiaries of the international oil companies supplied Parry Leon & Hayhoe, who then sold the oil to Genta, which in turn provided the Rhodesian subsidiaries of the oil companies with products for the local market. By March 1967 the South African Loop had finally been phased out, and from then on the Direct Route supplied virtually all of Rhodesia's oil.

During 1967 a total of 10,796 railway tank-cars were sent to Rhodesia on the Direct Route. The local subsidiaries of Shell and BP supplied forty-one per cent of the oil, followed by Mobil (eighteen per cent), Caltex (sixteen per cent), Sonarep (thirteen per cent), and Total (twelve per cent). These figures are particularly interesting, since (except for the inclusion of Sonarep) they are similar to the shares of the Rhodesian market held by the companies at the time of UDI. This suggests that there was considerable cooperation between the oil companies to preserve their stake in Rhodesia.

*　　*　　*　　*

Back in London, however, the British Government seemed to be far more interested in deflecting African demands for tougher action against the Smith regime than it was in discovering the facts about Rhodesia's oil supplies. Wilson claimed that the establishment of the Beira Patrol had been a severe blow against the rebel regime. At the end of April 1966, two weeks after the Royal Navy had begun patrolling the Mozambique Channel, the Prime Minister told Parliament that the oil embargo and 'the action we took recently at Beira' had been decisive in forcing the Smith regime to sue for peace.[8] Supposedly, then, as a result of Britain's tough stand British and Rhodesian representatives held 'talks about talks' to explore whether there might be a basis for more formal negotiations; these preliminary discussions, which opened in London on 9 May 1966, continued intermittently over the next three months. In the end, however, the 'talks about talks' broke down over Smith's unwillingness to offer sufficient concessions. This was to be only the first of a long series of unsuccessful attempts at negotiations.

The British Government was apparently unaware as yet that it was the international oil companies, particularly Shell and BP, which were the main sanctions-busters, and persisted in regarding South Africa and Portugal as the culprits. But what could be done? Wilson admitted to his Ministers: 'We can't afford to impose sanctions on both Portugal and South Africa.'[9] In August 1966 he told Cabinet colleagues about a new plan: 'If we can isolate Portugal, knock them hard on the head, this will stop the main South African leak into Rhodesia along the railway which crosses Portuguese territory.' But the plan – which one Minister described as Wilson's 'Portuguese gimmick' – was rejected by the Cabinet on the grounds that it might still lead to a confrontation with South Africa.

Wilson realised that he had to at least be *seen* taking some action to tighten sanctions. At the January 1966 Commonwealth summit conference, at which he had made his 'weeks not months' prediction, it had been decided to meet again in six months' time if the rebellion had not been crushed. The British Government had later suggested postponement until September, while the 'talks about the talks' were held. But these preliminary negotiations had now broken down, and the Commonwealth was once more angry at the failure of Britain's Rhodesian policy. When Wilson left the Cabinet meeting at which his proposal to tighten sanctions had been rejected, he was a worried man. 'My God,' he confided to Richard Crossman, 'you've left me with a handful of trouble.'[10]

Wilson later described the Commonwealth summit at Marlborough House as a 'nightmare conference'.[11] He opened the summit by apologising for his earlier prediction: more time would be needed, the British Prime Minister explained – although there were signs that sanctions were at last beginning to bite. The Cabinet met on 10 September 1966, half-way through the summit, and it was clear that major concessions would be needed if the conference was not to break down completely. Once more Wilson pressed the Cabinet to force Portugal to abide by sanctions and end the flow of oil from Lourenço Marques into Rhodesia.

'This is all irrelevant,' Crossman interrupted, 'because even if the oil does go by that route – by railway – there's an alternative route [through South Africa] at not much greater expense.' Crossman gleefully noted in his diary that it was 'the first time I've caught the PM out on a technical fact'.[12]

Crossman had correctly pointed out that oil could still be transported by the Beit Bridge road, which had been used in the early months of 1966. But this route had proved very expensive and would have considerably increased costs for Rhodesia. Action against Portugal would also have represented a signal to South Africa that Britain really meant business over sanctions. It is therefore unfortunate that Wilson's proposal to put pressure on Portugal was turned down by his Cabinet Ministers.

Two days later the Prime Minister again faced his angry Commonwealth colleagues across the table at Marlborough House. Wilson began by explaining that his Government would approach the Smith regime for further talks, and that if these failed, Britain would then go to the United Nations in December 1966 to propose mandatory sanctions, binding on all nations, on trade with Rhodesia. This would include a mandatory oil embargo against the rebel regime. Reluctantly the Commonwealth leaders agreed to the British proposals.

The subsequent negotiations with Smith took place in a dramatic confrontation aboard the Royal Navy's *Tiger* on

2 December 1966. Agreement was reached on a settlement which would have led to the resignation of the Smith regime, followed by an interim period of direct British rule pending the election of a new Government. Majority rule might have been achieved within fifteen years. But following the arduous negotiations, Smith hedged, claiming he would have to fly back to Salisbury to consult his Cabinet. After his return, the rebel regime decided to reject the *Tiger* terms.

A few days later Britain implemented her pledge to the Commonwealth leaders. George Brown, who had now become Foreign Secretary, flew to New York to address the UN Security Council on 8 December 1966. In his speech he pointed out that sanctions on trade with Rhodesia were still voluntary, but that the time had now come to tighten the embargo. For the first time since the Ethiopian invasion of 1935, an international body was being asked to launch economic warfare. Brown proposed a mandatory ban on the purchase of Rhodesia's major exports (asbestos, iron, chrome, sugar, tobacco, copper, meat, and leather) and on the sale of arms. The Foreign Secretary acknowledged that Britain had already come under strong pressure to include oil in mandatory sanctions, and declared: 'If an amendment in this sense were to be made in acceptable terms, my delegation would not oppose it.'[13] He made it clear, however, that the UK would only accept a *mandatory* oil embargo as long as sanctions were not allowed 'to escalate into economic confrontation with third countries' – a clear indication of Britain's reluctance to take action against South Africa and Portugal. Ironically, it was on the very day that the Foreign Secretary addressed the United Nations that Shell and BP despatched their first direct consignment of twenty tank-cars of oil to Rhodesia out in Lourenço Marques.

Following a lengthy debate, the Security Council finally put the sanctions resolution to the vote on 16 December 1966. The original British draft, which had not included any reference to oil, was approved by eleven votes to none, the four countries which had abstained on the Beira Resolution – France, the Soviet Union, Bulgaria, and Mali – doing so again, for similar reasons. Two amendments on oil were then considered, the first being a proposal requesting 'the United Kingdom to prevent by all means the transport to Southern Rhodesia of oil or oil products'. Britain's fear was that this motion would lead to a confrontation with South Africa and Portugal. But although seven countries voted in favour, and none against, there were eight abstentions (including the UK) and the amendment was therefore defeated. A weaker amendment merely imposing a mandatory ban on 'the supply of oil, or oil products, to Southern Rhodesia' was then approved by fourteen to none. Only France abstained, on the grounds that Rhodesia was not a United Nations responsibility.

In practice the new UN resolution did virtually nothing to tighten the oil embargo. Most UN members had already banned the export of oil to Rhodesia, and South Africa and Portugal were the only two major countries which had rejected the UN's earlier call for a voluntary embargo. According to Wilson, a *mandatory* embargo had not been introduced back at the time of UDI because of 'dangers that this might escalate into other resolutions'.[14] But if sanctions had been mandatory right from the start, before oil supplies had begun 'leaking' from Rhodesia's neighbours, then South Africa and Portugal might well have decided that they had little option but to implement the embargo. In leaving it until a full year later, and in accepting the mandatory embargo only reluctantly, Britain lost any influence it might have had over Pretoria and Lisbon.

*   *   *   *

Britain's subsequent pleadings with South Africa and Portugal fell on deaf ears. It was almost certainly known in Pretoria and in Lisbon – even if not in London – that the British-owned oil companies were supplying half of Rhodesia's oil. The South African and Portuguese Governments failed to see why they should be the ones to halt this trade. The fact that Britain had not been able to discover more than the most elementary facts about how Rhodesia was importing its oil prevented the Government from taking any effective action to enforce the embargo.

In January 1967 Herbert Bowden, who had replaced Arthur Bottomley as Commonwealth Secretary, was asked in Parliament whether he would 'give an assurance that a close and effective scrutiny is being kept by Great Britain upon all oil supplies passing into Rhodesia.'[15] Bowden

172

replied in the affirmative: 'A close scrutiny is kept.' A few months later Foreign Office Minister of State William Rodgers was asked about refined oil products entering Rhodesia. He merely replied that he would 'prefer to say nothing which might reveal the extent of our knowledge of the regime's complicated and expensive methods of procurement'.[16] On 29 June 1967 George Thomson wrote to Ben Whitaker, MP, to say that the Government was 'absolutely satisfied that British oil companies are not involved in the supply' of oil to Rhodesia. In this succession of statements MPs were therefore given the completely false impression that a close watch was being kept on the flow of oil, and that Shell and BP were not busting sanctions. The Parliamentary cover-up had begun in earnest.

In spite of what William Rodgers said, the Government apparently had remarkably little idea of the 'complicated' methods used by the Rhodesian regime to import its oil requirements. Yet anyone with a rudimentary knowledge of the oil industry and a fairly logical mind should have realised that right from the start there was always a strong likelihood that Shell Mozambique was involved in sanctions-busting. Determined efforts should therefore have been made to investigate this possibility.

After UDI Shell and BP continued to market about half of Rhodesia's oil, as was obvious from the fact that Shell and BP petrol stations inside Rhodesia still sold fuel under the trademarks of the British-owned companies. Simply by counting the number of railway tank-cars that passed along the line it could be confirmed that by the end of 1966 virtually all of Rhodesia's oil requirements were being railed directly from the port of Lourenço Marques into Rhodesia. Obviously, the Sonarep refinery at Lourenço Marques was a possible source of part of the oil sent to Rhodesia; but after existing commitments to supply the internal market in Mozambique had been met, with some additional exports to South Africa and Swaziland, the refinery only had available an excess of about 1,500 barrels a day for possible sale to Rhodesia. This would have accounted for less than one-fifth of Rhodesia's requirements. Most of the oil supplied to Rhodesia was therefore landed in refined form at the port of Lourenço Marques. Shipping movements are relatively easy to monitor, and it should therefore have been clear that

Shell Mozambique handled almost half of the oil products shipped to the port. There were therefore strong grounds for believing that Shell and BP oil could be reaching Rhodesia.

The next question was whether Shell Mozambique was *aware* that much of the oil it handled was probably destined for Rhodesia. Even the most cursory investigation into the sales of Shell Mozambique would have shown that one customer purchased far more oil than anyone else: this was a forwarding firm known as Parry Leon & Hayhoe. The quantities it purchased actually amounted to around forty per cent of *all* the oil products landed at Lourenço Marques for Shell Mozambique. Sales to Parry Leon & Hayhoe were made in 'bond' – that is, without customs duty being paid – which meant that the oil must have been for export from Mozambique. Records would also have revealed that Parry Leon & Hayhoe only began bulk purchases of oil after UDI. The quantity involved represented about half of Rhodesia's entire requirements. A process of logical deduction should therefore have suggested that Shell Mozambique was the major source of Rhodesia's oil. This should also have been appreciated by the Directors of Shell Mozambique, and the company's auditors (Price Waterhouse).

In fact, much of this deductive work should have been unnecessary. For British diplomatic representatives were actually *told* that Shell and BP were busting sanctions. Jardim claims that in 1966 both he and the Portuguese Foreign Minister, Dr Franco Nogueira, had confronted the British Ambassador in Lisbon, Sir Archibald Ross, with what he described as 'conclusive proof' about the role of Shell and BP in supplying Rhodesia. Sir Archibald, Jardim later recalled, 'was flabbergasted and worried'.[17] On another occasion José Teixeira, head of Shell Mozambique, had explained to a member of the British Embassy in Lisbon that some of the international oil companies were already sending direct supplies to Rhodesia. At this stage Teixeira apparently 'admitted having helped the Rhodesians by pointing out the loopholes in the embargo and he regretted that Shell (Mozambique) had not been involved in the trade' (B7.23). It was only a few days before Shell Mozambique did indeed join the other companies in directly supplying the rebel regime.

The British Consul in Lourenço Marques, Henry Downing, was particularly well placed to discover how oil was reaching Rhodesia, and Jardim, who recalled his 'always pleasant social contacts' with the diplomat, would have been an obvious source of information.[18] Downing told Bingham that he had first learned that direct oil shipments were being made from Lourenço Marques early in 1966, although Bingham commented that Downing had 'no hard evidence' as to who was sending the oil (B7.17). Bingham's phraseology does suggest, however, that the Consul may have had his suspicions. But the fact is that as time went by British diplomats appeared to show less and less interest in pursuing the sanctions trail. 'One got a little blasé about it,' Downing admitted to Bingham. 'One knew one was not going to get anything' (B8.80).

During 1967 there was a stream of accusations from well placed sources about the involvement of the international oil companies. The Portuguese Government was still anxious to show that it was the oil companies – not Portugal – which were responsible for the failure of the embargo, and in February 1967 the Portuguese Foreign Minister, Dr Franco Nogueira, again told the British Ambassador in Lisbon that his Government possessed damaging evidence about the involvement of Western companies in sanctions-busting (B6.11). Three months later Dr Nogueira was more explicit. During the first quarter of 1967, he said, eighty-five per cent of Rhodesia's oil was provided by British and American companies – Shell being the largest supplier.

The Portuguese Government then gave the French Foreign Ministry figures showing the precise number of railway tank-cars sent by each of the Southern African subsidiaries of the oil companies to Rhodesia. Shell/BP with 430 tank-cars, supplied almost half of the 866 tank-cars despatched in April 1967. Caltex (164), Total (145), and Mobil (127) provided the remainder. Although this information was passed on to the British Government at a meeting at the Quai d'Orsay on 7 June 1967, Bingham claims that 'the British authorities roundly disbelieved these figures' (B7.55). According to H. McNeil, one of the officials monitoring the oil embargo at the Ministry of Power, the statistics had been either faked or specially selected to implicate the British oil companies.

Accusations also came from other sources. In May 1967 President Kaunda argued that the British Government was deliberately permitting the shipment of oil to Lourenço Marques because it feared a confrontation with South Africa if any further attempts were made to block this supply. Shortly afterwards Rowland told James Bottomley, a civil servant at the Commonwealth Relations Office, that sixty per cent of Rhodesia's oil was being supplied by Shell and BP.

Why was it then, that the British Government made so little effort to check these reports? Between May 1966 – shortly after the Beira Resolution – and January 1967, there were apparently no substantive meetings with Shell and BP to discuss the oil embargo. Yet this was the critical period during which direct shipments to Rhodesia began. Even during 1967, when there was more regular contact between the Government and the oil companies, the discussions were mainly confined to considering suggestions for cutting off the flow of oil from Portuguese and French oil companies. Little attempt was made to inquire whether the British firms might be involved.

During the first two years of the oil embargo, the British Government appears to have unquestioningly accepted assurances from the UK oil companies that they were not busting sanctions. The London offices of Shell and BP in turn merely repeated 'assurances' received from Louis Walker, the head of their operations in South Africa and Mozambique. On 5 June 1967, for instance, Walker had cabled to Dirk de Bruyne at Shell Centre to 'give you my categorical assurance that except for very minor transgressions by individuals in comparatively junior positions in the first days of the Rhodesian affair all group companies in South Africa and Mozambique have rigidly adhered to your instructions and the relevant Orders in Council regarding the supply of products to Rhodesia' (B1.59).

Yet the fact was that Shell and BP's subsidiaries in Southern Africa had become deeply involved in sanctions-busting within a few months of the introduction of the oil embargo.

By the midle of 1966, Bingham explains, Walker was 'no longer willing officiously to strive to comply with the spirit of the [UK Sanctions] Order' (B7.23). One of the first portents of Walker's attitude was in a 'strictly private and confidential' letter which he wrote to Dirk de Bruyne at Shell

176

Centre in London on 2 June 1966. Walker admitted that 'the "oil boycott" has collapsed completely and is now no more than a diseconomy and an inconvenience to the Rhodesians' (B7.24). The tone of his letter suggested unmistakably that he intended to make little effort to prevent Shell oil from reaching Rhodesia. Why, therefore, did de Bruyne not immediately reply to Walker, stressing the need to enforce sanctions more strictly?

Bingham has described it as 'unfortunate' that Walker gave 'categorical assurances' to his head office in London that Shell Mozambique was *not* involved in sanctions-busting, when he knew that the company was in fact providing half of Rhodesia's oil (B14.13). The most likely explanation for Walker's actions is that he believed that sanctions would never work, and so failed to see why Shell and BP should lose their substantial stake in the Rhodesian market. But 'the easiest course for Mr Walker personally', Bingham added, 'would have been to disclose his knowledge and suspicions to Shell Centre and let others bear the burden of deciding what to do' (B14.14).

Shell told Bingham that the main reason why head office had not questioned Walker's assurances was that it would have been entirely alien to its general relationship with local subsidiaries. Instructions on policy had been given, and implementation was a matter for the local General Manager. This argument has a certain applicability with regard to Shell South Africa, which was registered in Cape Town. But the position of Shell Mozambique, which was incorporated in London, was rather different. As a British company it was obliged to respect the UK Sanctions Order, and its Directors therefore had a special obligation to see that the law was being upheld. The fact that the Rhodesian embargo was obviously being breached should only have encouraged the head offices of Shell and BP to ensure that their local subsidiaries were in fact respecting sanctions.

Surely then, the head offices of Shell and BP should have carried out their own independent investigation in Southern Africa? Although in South Africa, it is true, the Official Secrets Act might have made it difficult to investigate the local oil companies, a Shell team sent to South Africa in February 1968 apparently did *not* find its inquiries seriously hampered by the Act.

In spite of the fact that there was no similar official secrets legislation in Mozambique, Walker told Bingham that it would still have been very difficult to have concluded an investigation in Lourenço Marques:

> It would have been extremely risky to have tried to do a bit of amateur espionage. The country was on a war footing . . . There was at one stage talk of a Shell man, one of our people, trying to gain that sort of information. London sounded me out, and I said . . . it wasn't a proper thing to ask any of our staff to do that. Very nasty things could happen to you in Mozambique if you stuck your nose into what the authorities regarded as security matters (B6.6).

Since Walker knew that in fact Shell Mozambique *was* involved in sanctions-busting he may have been exaggerating these difficulties in order to deter an investigation by head office. Jardim has contradicted Walker's assessment of the situation in Mozambique: if an official had tried to discover who was supplying Rhodesia's oil, he claims that he personally would 'have invited him in for a meal'.[19] Jardim's subsequent role in exposing the sanctions scandal, and the fact that Portugal had at the time provided Western Governments with several reports on oil traffic to Rhodesia, suggests that insurmountable barriers might not have been put in the way of Shell/BP investigations in Mozambique. Nor would such an inquiry have involved a 'risky' exercise in 'amateur espionage', since the sales records of Shell Mozambique would have revealed the magnitude of supplies to Parry Leon & Hayhoe. The local head of Shell Mozambique – José Teixeira (and later João Vasconcelos) – would then have been able to confirm to London representatives that the company was indeed supplying Rhodesia through this intermediary.

Shell and BP executives from London visited South Africa and Mozambique from time to time on general business, and could have taken the opportunity to verify Walker's assurances; Shell officials also continued to visit Rhodesia after UDI: Dirk de Bruyne went to Salisbury in November 1966, and he was followed one year later by John Francis. But it should even have been possible to discover the facts without ever leaving London. The General Manager of

Shell/BP Rhodesia (G. L. Whitehead and later Robert Barrie) normally visited London once a year. Louis Walker, the head of Shell South Africa, came to London a total of forty-five times after UDI. On a number of occasions Walker even called at the Ministry of Power to discuss sanctions with Barry Powell (B7.38).

Visits by Shell executives from South Africa would surely have provided an occasion to inquire into the possibility that the company's Southern African subsidiaries were involved in sanctions-busting. Indeed, it is quite possible that Walker did informally tell his London management about trade with Rhodesia. Certainly he must have given *some* explanation to British Government officials when the two sides met at the Ministry of Power on 30 May 1967. Walker later admitted that on this occasion he had hedged in his replies to Barry Powell. But, he claimed, 'I felt that I had told them enough for them to realise what the situation was' (B7.38). If Walker actually hinted this much to British Government officials, it is difficult to believe that he would not have told at least as much to the Directors of his own company. Indeed, it is quite possible that Walker told them rather more and rather earlier.

Because of the frequent contact between oil company executives from London and Southern Africa it should not have been impossible to discover the truth about sanctions-busting. There seem to be only two explanations for the apparent ignorance of head office staff. It is possible that some Shell and BP executives in London *did* know that their Southern African subsidiaries were involved in trading with Rhodesia but that they chose to take no action. Alternatively, they simply failed to ask some very obvious questions – possibly on the grounds that they did not want to be 'burdened' with potentially embarrassing information. In either case, one cannot help noticing the poor upward flow of information within the oil companies – both from Southern Africa to London, and then within the hierarchy at head office.

*   *   *   *

The British Government, in turn, seems to have been extremely negligent in unquestioningly accepting the 'assurances' it received from the head offices of Shell and BP. As late as 1967 the UK seemed to be completely confident that the Southern African subsidiaries of Shell and BP were innocent. Rhodesia, it was believed, was obtaining part of its oil requirements from the Sonarep refinery, and the remainder must be 'leaking' from refined oil supplies supposedly sent from Lourenço Marques to South African customers. The Government assumed that these leaks were probably emanating from the non-British companies, such as Total, and that if oil was reaching Rhodesia, obviously foreigners were involved; British companies were respectable firms, after all.

France was singled out as a major villain. On 19 June 1967, during Wilson's visit to Paris, the British Prime Minister accused Total of defying the oil embargo. Wilson later recalled that he told President Charles de Gaulle 'to clamp down on sanctions-breaking by French oil interests sending oil to Rhodesia through Mozambique'.[20] De Gaulle promised to study the question, but pointed out that the anti-colonial wars faced the Portuguese with great difficulties in Angola and Mozambique and that France tried on the whole 'not to make life too difficult' for Portugal.

Wilson continued to blame the French. On 11 December 1967 he told Parliament: 'The bulk of Rhodesia's oil supplies comes from Lourenço Marques, and the Compagnie Française des Pétroles [Total] supplies Lourenço Marques both with large quantities of crude oil for the refinery and of refined products.'[21] The British Government continued to discount Portuguese Government reports that Total was a relatively minor supplier of oil to Rhodesia and that Shell and BP were providing half of Rhodesia's requirements.

Since Rhodesia was regarded primarily as a British responsibility, and the UK was under strong international pressure to achieve a solution, any suggestion that British companies were breaking sanctions was particularly damaging. The British Government therefore began to search for a means of stopping what they believed to be the 'leakage' to Rhodesia – or at least showing that British companies were not involved. Following the breakdown of the *Tiger* talks, and the subsequent UN decision to impose mandatory sanctions in December 1966, the UK investigated a series of initiatives to achieve these goals. During the year four different schemes were considered.

On 12 January 1967 George Brown summoned the British oil companies to the Foreign Office. Bingham reports that the Government needed 'to keep control of the Rhodesian situation', and that oil sanctions had to be made sufficiently effective 'to avoid either pressure for enforcement measures against South Africa and Portugal, or action by oil-producing states to deny the oil companies the right to export oil to Southern Africa' (B6.16). George Brown went on to propose the Government's first initiative: that oil shipments to Mozambique should be rationed so as to limit supplies available for onward sale to Rhodesia. The Foreign Secretary claimed that the Portuguese authorities might be persuaded to cut off supplies to Rhodesia in order to protect their own oil imports; would the oil companies be willing, if necessary, to restrict or ration Mozambique's supplies from overseas? Shell and BP responded that there were numerous difficulties, and they doubted that such a scheme would be effective.

Six days later a follow-up meeting was held at the Ministry of Power, at which Angus Beckett, head of the Petroleum Department, and Barry Powell both expressed sympathy with the oil companies' point of view. Shell's note of the proceedings records that the Ministry of Power had found 'that the Foreign Office, because of political necessity, insisted that something should be done to stop the oil supplies to Rhodesia' (B/A 240); but according to the note, the Ministry of Power 'had always advised them that limitation of such supplies was not practical, nor would it be effective, and that furthermore it could not be expected that the British oil companies would participate in any such scheme on a voluntary basis. Beckett and Powell therefore went on to suggest a further meeting with the Foreign Office. Dirk de Bruyne of Shell noted that 'the Ministry of Power would try to be represented at the meeting in order to add their views to ours' (B/A 241).

Dr Jeremy Bray, who was then Parliamentary Secretary at the Ministry of Power has claimed that his civil servants 'advised the oil companies on how they could best undermine the Foreign Secretary's plans'.[22] Dr Bray told the House of Commons that Beckett and Powell had suggested a further meeting with Foreign Office Minister of State George Thomson, 'whom they regarded as more flexible', or Sir Paul Gore-Booth, Permanent Under-Secretary at the Foreign Office. Ministry of Power officials would then try to be present 'to support the oil companies' views'. The meeting actually took place on 24 January 1967, and in the end it was at a more junior level, with Rose and Robert Farquharson of the Foreign Office. Bingham remarked that 'little new ground was covered' (B6.19).

On 8 March 1967 *all* the international oil companies operating in Southern Africa were invited to the Ministry of Power in London to discuss further the coordination of the possible rationing scheme to reduce supplies to Mozambique – the first proposal. Shell and BP both attended. However, the four American companies which were represented (Mobil, Caltex, Esso, and Aminoil) were reluctant to provide data on their shipments to Lourenço Marques, and the French firm of Total refused even to attend the meeting. The British Government therefore abandoned any further consideration of the scheme.

The second scheme put forward by the British Government was based on the quite incorrect assumption that the Sonarep refinery in Lourenço Marques was the main source of Rhodesia's oil. The proposal was that Shell and BP should purchase all output of the Sonarep refinery not required for the internal market in Mozambique. This excess oil, it was hoped, would therefore not find its way to Rhodesia. George Thomson, then Minister of State at the Foreign Office, put this scheme to the oil companies at a meeting on 13 April 1967. Shell and BP replied that it might cost the two companies £1 million a year. Why, the oil companies asked, could not the British Government itself purchase this oil? Thomson later recalled that when he put this proposal to the Treasury he 'got a very dusty response' (B6.21).

The third proposal got closer to the real problem – it involved an attempt to prevent oil supplies which were in transit from Lourenço Marques to South Africa from being 'diverted' to Rhodesia. Traditionally part of the South African market close to the Mozambique border had been supplied from the port of Lourenço Marques in order to minimise transport costs. But it became increasingly clear that shipments of oil which were nominally destined for South Africa might well be reaching Rhodesia. The British Government therefore approached Shell and BP to ask whether they could re-route these transit supplies from Lourenço Marques to Durban.

This proposal was first put to the oil companies in a letter from the Ministry of Power on 12 October 1967. The Government did not believe that re-routing Shell and BP supplies would actually cut off Rhodesia's oil, since it was thought (quite incorrectly) that this mainly originated from the French and Portuguese oil companies – but it would at least demonstrate that British oil was not reaching Rhodesia, and enable diplomatic pressure on France and Portugal to be intensified. Shell and BP, however, argued that the proposal to divert supplies from Lourenço Marques to Durban would considerably increase transport costs, and the scheme was quickly dropped (B6.22).

The final proposal was that the oil companies should end the system whereby some South African bulk customers purchased supplies at Lourenço Marques and then made their own transport arrangements. Instead, the British Government suggested, Shell and BP should send the oil to their own depots inside South Africa, where it could then be sold to bulk customers. This scheme was raised with the oil companies by George Thomson, who had become Commonwealth Secretary, and Foreign Office Minister of State Goronwy Roberts on 11 December 1967. Again, however, both Shell and BP raised problems, claiming that there would be insurmountable difficulties in changing their marketing arrangements. The fourth proposal was therefore killed (B6.30).

Thus, during 1967, four plans to tighten sanctions had been considered and rejected; but they had mainly been based on a fundamental misconception: that it was not Shell and BP which were busting sanctions, but the French and Portuguese companies. It was also unfortunate that the schemes were not raised until over a year after the oil embargo had been introduced, since the slow pace of the Government's reaction to the 'leakage' from Lourenço Marques suggested that Britain was not really serious about sanctions. Even when moves were finally afoot Shell and BP repeatedly raised objections and claimed that the proposals were im-

183

practicable, frequently with the backing of the Ministry of Power – which generally seemed to reflect their interests. Because the Ministry of Power had technical knowledge, the Commonwealth Relations Office and the Foreign Office usually bowed to its advice.

By the end of 1967 Britain's sanctions policy had reached a new level of hypocrisy. During the Commonwealth Secretary's meeting with the oil companies on 11 December 1967, Thomson had said that 'while he would be most interested to hear their views on what more could be done to stop this leakage [from Mozambique to Rhodesia], he was more concerned at this meeting to discuss how current Portuguese propaganda aimed at pinning the main responsibility for the leakage through Mozambique on British and American companies – especially Shell – could most effectively be countered' (B6.27). But just the day after making this astonishing statement, Thomson again tried to lay the blame for the failure of sanctions on other members of the UN. There was a need for international action, he told Parliament, 'to make sure that all the countries which voted for the mandatory resolutions are taking all practicable steps to ensure that firms under their jurisdiction honour those resolutions.'[23]

* * * *

Already by late 1967, however, the British Government was receiving increasing evidence of a most embarrassing discovery. It was becoming obvious, even to the British Government, that Shell and BP were actually supplying half of Rhodesia's oil. Although reports of sanctions-busting from foreigners, such as the Portuguese, had always been discounted, Robert Clinton-Thomas, the Foreign Office official responsible for monitoring sanctions, realised that there was growing evidence that British oil companies were indeed involved, and wondered if perhaps a newspaper exposé might force the politicians to face facts.

In August 1967 Clinton-Thomas encouraged *Sunday Times* reporter Colin Simpson to go out to Southern Africa to investigate.[24] Obviously Simpson's passport, which discribed him as a journalist, would only prove a liability. Within a matter of hours arrangements were therefore made for him to be issued with a new passport, and for the purposes of the trip he became a 'Company Director dealing in second-hand printing machinery'.

Lourenço Marques was Simpson's first stop. There a Foreign Office 'contact' fixed him up with an overnight stay on the sea-front – not in a luxury hotel, but in the attic of a warehouse near the port. It was already dark, but Simpson quickly realised that he was not alone. A few yards away he made out the figure of a man. The stranger came over and introduced himself as an assistant US Military Attaché based in an adjacent East African state. The American was also investigating the oil flow, he told Simpson, and showed the British journalist his sophisticated night vision equipment and a camera with a powerful telephoto lens.

The US Attaché explained to Simpson that he had made a financial 'arrangement' with railway officials to photograph the oil tank-cars. Early the following morning, at first light, Simpson and the American took pictures of the despatch labels contained in wire cages on the side of each oil tank-car, and with the help of a small bribe, the two observers were able to travel in the guard's van of a train that was carrying the oil into Rhodesia. At a small station thirty miles from Lourenço Marques, Simpson and his companion became nervous, and left the train, quickly hitching a lift back to Lourenço Marques.

From the Mozambican capital, Simpson flew to Umtali, to meet the train on its arrival. Again he photographed the oil tank-cars, confirming that they were the same ones he had seen at Lourenço Marques – evidence, he believed, that Shell and BP were probably supplying Rhodesia. Simpson then went on to Salisbury, where he met John Hennings, the head of the British mission in the Rhodesian capital, to outline his findings. Back in London, Simpson gave a written report to Robert Clinton-Thomas and James Bottomley at the Foreign Office.

Part of Simpson's story appeared in a *Sunday Times* Insight exposure of the sanctions-busters. On 27 August the *Sunday Times* carried a report that Shell/BP South Africa was supplying 'independent dealers' who were selling oil to Genta, with Shell/BP Rhodesia then purchasing the oil from Genta. This seems to have been the first published account of the paper-chase. A week later the *Sunday Times* reported that there had been 'polite murmurings at the highest executive level of Shell' that it was 'unfair' to single out the Anglo-Dutch company.[25] A Shell spokesman did, however, explain that London-based executives in Southern Africa delicately avoided inquiring where their oil was going. 'Asking a lot of questions wouldn't exactly be well received out there, you know.'

The Government also appears to have confirmed Simpson's story from its own intelligence sources. Bingham admitted that by the end of October 1967 officials were investigating 'the possibility that the oil companies' products were being diverted in Mozambique after delivery to the customer for [purported] carriage to the Transvaal' (B6.23). British intelligence sources in Mozambique apparently followed a number of railway tank-cars containing oil sold by Shell and BP to South African customers (including Parry Leon & Hayhoe). Some of these tank-cars, however, never crossed into South Africa – and were presumably sent to Rhodesia.

The British Government then gave the head offices of Shell and BP several lists of numbers of railway tank-cars which had gone in what was euphemistically described as the 'wrong' direction. But initially these reports were not taken too seriously. A Shell note on 13 February 1968 records that H. McNeil of the Ministry of Power 'handed us a further list of recent tank-car "sightings"' and that another official, Barry Powell, then asked Shell simply to treat it as 'information given to us for what it was worth'. John Francis of Shell added that Powell 'did not ask that we should return to him any comments thereon, nor did he ask for any further views on the earlier and similar list'. Francis added, 'Naturally, I did not volunteer any' (B/A 252).

A row over sanctions-busting was the last thing that the oil companies wanted at this particular moment. By the end of 1967 their position in South Africa was at a particularly critical juncture since the South African Government was becoming increasingly uneasy over its dependence on foreign oil companies, which, it was believed, could easily be subject to outside control – particularly by the British Government. Commercially, the outlook for the oil industry in South Africa was bright: oil consumption was increasing rapidly, and a new refinery was on the drawing board. But

Shell and BP were worried that their rivals would seize any opportunity to reduce the British firm's dominant position in the lucrative South African market.

Shell and BP tried to fight back by each purchasing a seventeen and a half per cent shareholding in Trek, a South African oil company. 'Our commercial relationships in South Africa are at an extremely delicate stage,' a Shell note explained in February 1968, and participation in Trek 'will give us a good chance of obtaining a foothold in the projected Government refinery'. They were also hoping to secure for themselves 'a fair share of the impending Government tender for emergency crude supplies' (B/A 250).

BP later told Bingham that 'in these circumstances it would have been commercially damaging' for the company's South African subsidiary 'to be seen to respond to the pressure of their English shareholders' over Rhodesia.[26] (In the end, however, Trek lost the contract to participate in the new refinery, and Total and the National Iranian Oil Corporation took up a minority shareholding in the project.)

*   *   *   *

Disquieting evidence was now accumulating from different sources to suggest that Shell and BP oil was reaching Rhodesia in massive quantities. By the end of December 1967, Dirk de Bruyne recalled, Shell had become 'a bit uneasy about the whole thing' (B6.41). On 28 December 1967 senior Shell and BP officials discussed what had become now their 'disquiet about the situation in regard to supplies to Rhodesia'.[27] This was to be the first in a long series of delicate discussions that the oil companies were to hold among themselves over the next few weeks.

Unease and disquiet rapidly developed into panic. Jack Miller, the General Manager of Shell's marketing company in South Africa, visited London during the first few weeks of the New Year, and apparently told Hugh Feetham at Shell Centre that customers, or *a* customer, was obtaining supplies for Rhodesia from Shell Mozambique (B6.42). Feetham later recalled that Parry Leon & Hayhoe had probably been mentioned as the customer concerned. Details of Jack Miller's enormously embarrassing disclosure were recorded in a confidential Shell note on 22 January 1968: 'Shell Mozambique has reason to believe that the petroleum [it markets] is being diverted to Rhodesia' (B/A 242).

Over the next month the oil companies held a series of top-level meetings to decide what to do. Initially it was suggested that a Board meeting of Shell Mozambique should be called. The Directors of Shell Mozambique who were British citizens, and therefore covered by the UK Sanctions Order, were then John Francis and William Fraser. But since some of the Directors were abroad at the time, it was proposed that the joint Shell/BP parent company, Consolidated Petroleum, should meet. Then, on 23 January 1968, just the day before the Consolidated Board was due to convene, the meeting was cancelled 'because of the difficulties of implicating the Directors in any report that they submitted to the Board' (B6.48), it being apparently feared that a formal meeting of the Consolidated Directors might put incriminating information on-the-record. The informal discussions continued.

BP then proposed that Louis Walker should be relieved of his post as General Manager of Shell Mozambique. Walker's conflicting responsibilities required him to be based in South Africa, as head of Shell's operations there, but at the same time made him responsible for seeing that British sanctions law was respected by Shell Mozambique. It would therefore be better, BP argued, if a UK citizen was appointed to head the company in Mozambique. However, Sandford recalled, 'Shell just wouldn't listen to it, it is as simple as that. They insisted on retaining him' (B6.50). According to Bingham, 'Shell undoubtedly had valid reasons for declining to make any change, since Mozambique had traditionally been regarded as an integral part of the South African operation and it was felt that the removal of Walker at that stage might well be construed as provocative by the Vorster Government in South Africa' But since this change in management concerned a British-registered company operating in Mozambique, it is difficult to see why this should have been regarded as 'provocative' by South Africa.

Shell also argued that Walker's removal would put them 'in a very awkward position, since there was no expatriate to take over as General Manager of Shell Mozambique'

(B/A 247). It seems surprising, however, that the Shell Group – which then employed over 170,000 people around the world – was unable to find another Manager for Mozambique. Shell concluded that Walker was 'well placed to look after our interests and to fulfil our responsibilities in Mozambique' – a very damaging admission on the face of it, since it suggests that Shell Centre may to some extent have been condoning Shell Mozambique's activities.

There then followed a bizarre exchange of messages between London and Cape Town. On 26 January 1968 Dirk de Bruyne wrote to Walker to request an assurance that Shell Mozambique was *not* breaking UK sanctions legislation. Walker responded positively three days later: 'We have not despatched any products at any time in contravention of the relevant Orders in Council' (B6.51). This categorical statement was soon discovered to be untrue, for on 15 February Arthur Sandford of BP admitted in a confidential note that 'Walker's cable to London assuring us that the Orders in Council had not been contravened was not correct in the light of the facts as we now know them . . . Shell Mozambique had reasonable cause to believe that the supplies . . . were going to Rhodesia' (B6.62).

De Bruyne's request for an 'assurance' appears to have been carefully contrived to provide documentary cover to guard the Directors of Shell Mozambique against charges that they had *known* about shipments to Rhodesia. The main evidence supporting this thesis is a draft letter, addressed to Walker by Shell executive Hugh Feetham, drafted three days before de Bruyne's request for an assurance was posted. Feetham's letter, which was never actually sent, admitted: 'We ourselves are inclined, although we did not say so in the letter [from de Bruyne], to think that the arrangements which you have entered into in South Africa and Mozambique are in all probability such as to place Shell Mozambique and its Directors in a position of liability to prosecution' (B/A 244). In the light of this frank admission, it looks as if de Bruyne chose his words with special care when, at the end of his subsequent letter to Walker, he asked him for an assurance that Shell Mozambique was *not* contravening the UK Sanctions Order.

Frank McFadzean, Shell's Managing Director responsible for Africa, later tried to explain this curious episode. He told Bingham that he took Walker's 'assurance' to be 'a very legalistic interpretation' – on the grounds that Shell Mozambique sold to Parry Leon & Hayhoe, not to an actual Rhodesian customer. 'My reaction at the time,' McFadzean added, 'must have been that this is squeezing it a little bit hard' (B6.52).

By the beginning of February 1968, however, the oil companies had decided that they had to face the unpleasant truth. More than two years after the introduction of the Sanctions Order, Shell sent out its own investigation team to Southern Africa. On 4 February 1968 Hugh Feetham and John Wilkinson met Louis Walker in Johannesburg, and then flew on for a short day-trip to Lourenço Marques. No written record was made of their investigation – probably because this would have put incriminating evidence on paper – but Wilkinson later admitted to Bingham that the visit had confirmed that 'there was indeed good ground for the suspicion that goods were going direct to Rhodesia' (B6.55)

Bingham commented that he was 'surprised that the [verbal] report made by that team did not cause some dissatisfaction with the information previously supplied from South Africa', but had nowhere heard that this was the case (B14.16). It is also interesting that the oil companies' concern seemed to be solely confined to the question of whether Shell Mozambique was busting sanctions (because of its legal obligation to abide by the UK Sanctions Order); the fact that the South African subsidiaries of Shell and BP were *also* involved in supplying Rhodesia caused little surprise – which suggests that this may have already been known to the head offices of the companies in London.

'At this point,' Bingham explains, 'something of a tussle developed between BP and Shell as to when the Minister should be told of the new situation' (B6.63). William Fraser, the BP Managing Director responsible for Africa, was conscious of the Government's shareholding in the company, and therefore insisted that the facts should be disclosed without delay: 'If it ever came into the open that BP had condoned such a thing [*i.e.* sanctions-busting] without informing the Government, it was perfectly obvious that the Government would in one way or another be forced to take control of BP.'[28] Surprisingly, however, neither of the two

Government representatives on the BP Board at the time – Sir Frederic Harmer and Lord Trevelyan – appear to have been involved in any of the discussions over sanctions-busting. This again suggested that the Government and its representatives on the BP Board had lost interest in the subject.

Shell was in a rather different position. As the company managing the joint Shell/BP companies in Southern Africa, it had the primary responsibility. But at the same time its status as a wholly private company, without any Government ownership, enabled it to act according to purely commercial criteria. Senior Shell Directors believed that the Commonwealth Secretary should be told about the embarrassing discovery, but as Bingham points out, they 'were much against disclosure until the full facts were known and until the companies could inform him of their proposed solution' (B6.63). Bingham adds, however, that other Shell executives went so far as to query 'the wisdom of officially putting HMG on notice of the action which was being taken'.

On 13 February 1968 John Francis of Shell met Barry Powell at the Ministry of Power to discuss sanctions. Francis failed to tell Powell that the Shell investigation team had just reported back from Southern Africa that Shell Mozambique *was* indeed handling supplies for Rhodesia. Instead Francis chose to pass on the 'assurance' which had been received a little earlier from Louis Walker claiming that the company was *not* contravening the Sanctions Order. Bingham later reported that 'Francis had not felt there was any particular need to burden Powell with the facts in detail, provided the companies had arranged matters so as to produce the end result he was seeking' (B6.60). 'Francis,' he added, 'had real doubts whether HMG should be put officially on notice of the action being taken' (B6.66).

Shell and BP continued this debate. On 19 February 1968 BP took the stand that 'a full disclosure of the facts should now be made to the Commonwealth Secretary' while Shell argued that 'quite a different approach was called for' (B6.68). BP's record of the discussion adds that 'Shell resisted point by point'. John Francis later admitted to Bingham that Shell was not all that anxious to expose itself to 'the undesirable consequences' of telling the Government that the company might be breaking the law (B6.70), and confirmed that BP 'was for the maximum amount of disclosure' and that Shell was, 'on the whole, not'.

The last straw, which forced the oil companies to admit the facts to the British Government, was a public outburst from Zambia. On 15 February 1968 President Kaunda held a press conference in which he accused Britain of being among the 'gangster nations' which were breaking the Rhodesian oil embargo.[29] He went on to claim that between January 1966 and August 1967 a total of 5,700,000 barrels of oil destined for Rhodesia had been delivered at Lourenço Marques. Among the tankers which had carried the oil were BP's *British Flag*, the *Caltex Cardiff*, and the *Mobil Energy*. Kaunda concluded: 'We have been told that sanctions were biting, but almost anything can bite – even a flea. What we have demanded of the British Government is for them to go back to their original statement that they would crush the rebellion, and not just fleabites.'[30]

The day after Kaunda's press conference it was reported that his charges had 'met with a cool reception in Whitehall,' since it was apparently felt that he had 'not produced any clear evidence' to back up his claims.[31] Shortly afterwards the Minister of State at the Board of Trade (Lord Brown) told Parliament that the information had been 'planted' on the Zambians, and they had 'most unfortunately, been taken in by it'.[32] The British Government apparently believed that it was Portugal which had 'foisted' these statistics on President Kaunda (B/A 268).

The Zambian President reported his allegations to the British High Commissioner, John Pumphrey. But as he later recalled: 'nobody took me seriously; the response was nothing, absolutely nothing, silence . . . '[33] It is true that Kaunda's allegations did not represent positive proof that Shell and BP oil was reaching Rhodesia, since the fact that refined products had been landed at Lourenço Marques did not necessarily mean that they were destined for Rhodesia. But the detailed evidence which he did produce should at least have been analysed more seriously by the British authorities. Shell and BP were therefore nervous that the Government would approach them to comment on the Zambian allegations. It was better to take the initiative, they decided, and to carefully break the news of the embarrassing discovery which had been revealed by their investigation in Southern Africa.

NOTES

1 *New York Times*, 12 April 1966.
2 Security Council debate, 17 May 1966.
3 Security Council debate, 17 May 1966.
4 Security Council debate, 18 May 1966.
5 Security Council debate, 18 May 1966.
6 Interview, 28 March 1979.
7 Jardim, *Sanctions Double-Cross*, p. 89.
8 Hansard, Commons, 27 April 1966, col. 710.
9 Richard Crossman, *Diaries of a Cabinet Minister*, vol ii, p. 18.
10 Richard Crossman, *Diaries of a Cabinet Minister*, vol ii, p. 20.
11 Wilson, *The Labour Government 1964–70*, p. 358.
12 Richard Crossman, *Diaries of a Cabinet Minister*, vol ii, p. 29.
13 Security Council debate, 8 December 1966.
14 Hansard, Commons, 20 December 1965, col. 1595.
15 Hansard, Commons, 27 January 1967, col. *389*.
16 Hansard, 17 July 1967, col. 1530.
17 *Observer*, 2 July 1978, and Jardim, *Sanctions Double-Cross*, p. 105.
18 Jardim, *Sanctions Double-Cross*, p. 44.
19 *Guardian*, 4 September 1978.
20 Wilson, *The Labour Government 1964–70*, p. 522.
21 Hansard, Commons, 11 December 1967, col. *48*.
22 Hansard, Commons, 7 November 1978, col. 792.
23 Hansard, Commons, 12 December 1967, col. 214.
24 This section based on letter from Simpson, 7 March 1979.
25 *Sunday Times*, 3 September 1967.
26 BP Submission to the Bingham Inquiry, 27 September 1977, p. 42.
27 File note prepared by R. Bishop, (BP), 1 January 1968 (unpublished).
28 Record note by William Fraser, 16 February 1968 (unpublished).
29 *Guardian*, 16 February 1968.
30 *Guardian*, 16 February 1968.
31 *Guardian*, 17 February 1968.
32 Hansard, Lords, 5 March 1968, col. 1220.
33 BBC Television, 19 September 1978.

13

10 The Government's Fingerprints

February 1968–February 1969

George Thomson's two meetings with the oil companies in
1968–69 mark a watershed in the sanctions story. For at
these meetings Shell and BP confessed that their jointly-
owned subsidiary in Mozambique had been busting sanctions, and explained that a swap arrangement had been set
up with Total for handling Rhodesian supplies. The significance of the two meetings, as one MP later explained, was
that the oil companies had succeeded in putting 'the Government's fingerprints' on the swap which Shell and BP had
made with Total, and which allowed them to continue to
supply their Rhodesian subsidiaries.[1]

On the afternoon of 21 February 1968 Shell and BP executives nervously trooped into Thomson's office, expecting
to find themselves under fierce attack. Already there had
been considerable tension between the two oil companies
over how much should be disclosed to the Government, and
neither Frank McFadzean (Shell) nor William Fraser (BP)
were looking forward to the discussions. On entering the
Commonwealth Secretary's room, they found Thomson
flanked by six officials from three different Ministries.

Two reports of the meeting are available – the official
Commonwealth Office record which has recently been 'de-
classified', and Shell's record. Bingham admitted that each
gave 'quite a different shape to the meeting,' but through
piecing together the evidence, it is possible to build up a
picture of the proceedings (B6.75). Thomson began with the
surprising statement that his immediate concern was not to
stop the flow of oil to Rhodesia through Lourenço Marques.
The Government, according to the official record, was
'under no illusions' about Britain's ability to end this leakage (B/A 256). McFadzean and Fraser must have breathed
a sigh of relief.

The official record then adds that the Commonwealth Secretary 'had been most discouraged to learn recently that, contrary to what the oil companies had led us to believe last year, there had in fact been a substantial leak of British oil through Mozambique to Rhodesia' (B/A 256). The Shell note recorded that Thomson had told African leaders during his visit to East Africa in November 1967 that 'no British company was supplying oil products to Rhodesia and . . . he had returned to London to discover that this was probably not correct' (B/A 260).

'Discouragement' seems an unusually mild reaction to such shattering news: for Thomson was referring to the fact that British oil companies had been providing half of Rhodesia's oil; and that Shell Mozambique had broken British law; that the head offices of Shell and BP had given incorrect assurances to the Government; and that the Commonwealth Secretary had then passed on these assurances to African Governments; the oil embargo had failed, and Britain's Rhodesian policy was therefore in tatters. It is perhaps surprising that the Commonwealth Secretary did not describe this news as anything less than disastrous.

Thomson concluded his introductory remarks by explaining that his immediate aim was to be able to say that no *British* oil was reaching Rhodesia. Any more positive action to tighten sanctions and cut off Rhodesia's oil, he claimed, would inevitably lead to a confrontation with South Africa – and the Government was simply not prepared to pay the price of alienating Pretoria. Britain was facing a severe economic crisis; the Pound had been devalued in November 1967; and its military forces were being withdrawn from East of Suez in order to cut public expenditure. Economic confrontation with South Africa was therefore out of the question: 'It was the considered view of the Cabinet that Britain, in those circumstances, should not add to its economic troubles by getting on a collision course with South Africa where there were such large British markets and investments.'[2]

The oil company representatives then confirmed that they had recently discovered that their jointly-owned subsidiary in Mozambique had been selling oil to 'suspicious' customers, who were probably diverting supplies to Rhodesia. They did not suggest, however, that these customers were in fact acting *on behalf of* Southern African subsidiaries of the oil companies. This could well have been a deliberate ploy by Shell and BP. Thomson, on his side, does not appear even to have raised the question of whether Shell Mozambique might have contravened the UK Sanctions Order, or whether prosecutions had been under consideration (see p. 267 for a discussion of Thomson's decision). The oil companies did assure the Commonwealth Secretary, however, that Shell Mozambique had now been removed from the chain of supply.

Shell and BP then went on to reveal the first details of the Total swap arrangement (see *Fig 5*). The Government's record of the meeting merely stated rather vaguely that oil for Rhodesia would now be 'drawn from non-British sources at Lourenço Marques' (B/A 256); but the oil companies' account of the discussion suggests that a more detailed picture of the new arrangement may have been given. The South African subsidiaries of Shell and BP, the Shell note reported, had probably initiated 'an exchange deal in Mozambique with some other company or companies not subject to British jurisdiction (Mobil, Caltex, CFP [Total])' (B/A 260). BP's Managing Director later amended the record, by adding that 'by the end of the conversation it was, I think, fairly clear to all concerned that CFP [Total] was the company most likely to be involved' (B/A 262).

Clearly the head offices of the oil companies told the Commonwealth Secretary rather less than they knew about the Total swap. A BP note, written by Arthur Sandford six days earlier, had spelt out in considerable detail how the new scheme operated:

> It is now proposed to eliminate Shell Mozambique, the UK registered company, from any involvement in the loading of supplies to this South African customer [Parry Leon & Hayhoe]. Shell South Africa, however, propose to continue the trade by arranging for CFP [Total] to load the RTCs [railway tank-cars] at the Lourenço Marques refinery with their product made available to us against exchange at Durban. To compensate CFP's effort Shell South Africa will assist them to supply their Transvaal depots (B6.59).

196

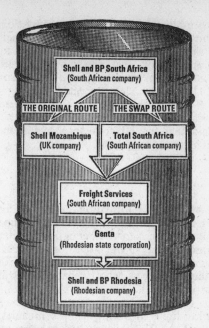

The Total swap

Fig. 5

The 'swap' arrangement used to supply Rhodesia between 1968 and 1971. Shell Mozambique was removed
from the supply chain. Instead the Shell/BP oil was
handled by Total South Africa at the Mozambican port
of Lourenço Marques.

The South African subsidiaries of Shell and BP were therefore to continue to supply the Rhodesian market. But in
order that their British-registered subsidiary in Mozambique
should no longer be involved, the oil was to be handled in
Lourenço Marques by Total. This arrangement was later
modified so that oil products supplied by Shell and BP South
Africa were handled by Total in Lourenço Marques for a
fee. In this way Shell and BP were 'laundering' their Rhodesian oil supplies.

BP later admitted that on 21 February 1968[3] Thomson had only been told 'something of what had happened'. Bingham points out that it was doubtful whether the Commonwealth Secretary appreciated that the new swap arrangement 'involved another oil company supplying suspicious customers at Lourenço Marques and receiving equivalent quantities of product from Shell/BP at convenient places in South Africa' (B6.75). Thomson himself later is reported to have said that the oil companies had been 'less than frank' with him.[4]

*   *   *   *

Ten years later, when information on the 21 February 1968 meeting was first publicly revealed, a row developed between Thomson and Wilson as to whether or not the Prime Minister had been informed about the discussions with Shell and BP. Even now, after so many of the participants have contributed to the 'who knew what' debate, it is still far from easy to sort out the facts.

Thomson claims that he reported to the Prime Minister in writing on his discussions with the oil companies. Certainly the minutes of the 21 February 1968 meeting were copied to Michael Palliser, Wilson's Private Secretary dealing with foreign affairs at Number Ten. George Brown has also confirmed Thomson's account. 'None of us then in office,' he said, 'could claim ignorance on the position.'[5] But Brown, who was Foreign Secretary, was particularly well placed to hear about the meeting with the oil companies, since Martin Le Quesne of the Foreign Office attended the meeting, and later received a copy of the minutes. Le Quesne would then presumably have informed his Minister of the outcome of the discussions.

One week after the meeting Thomson sent a detailed letter to Number Ten on the subject of Rhodesian oil. The letter was actually written by Thomson's Private Secretary (D. MacKilligin) to his counterpart at the Prime Minister's office (Michael Palliser). Unfortunately, however, it could hardly have arrived at Number Ten at a less auspicious moment. For on 15 March, the very day that the letter was received, Foreign Secretary George Brown resigned from the Government – the end of a turbulent relationship with the man who had defeated him for the Labour leadership in 1963.

Wilson later admitted that he did indeed see Thomson's letter, which he ticked in his usual style; but he claimed that it was not couched 'in the terms which have been suggested' by Thomson.[6] Although unfortunately only selected passages of Thomson's letter have been published, the extracts suggest that it was a confusing and ambiguous letter. The Commonwealth Secretary apparently admitted: 'We now know that a good deal of the oil which is getting to Rhodesia has been coming from refined products delivered to Lourenço Marques by French, British and US oil companies.'[7] He went on to specify that 'a good deal of the oil delivered to South African customers for (sic) Lourenço Marques, including some deliveries by Shell/BP, was being diverted within Mozambique to Rhodesia'.[8] These passages suggested that Shell Mozambique had indeed been handling oil for Rhodesia, even if unintentionally. The Prime Minister should therefore have realised that something had gone badly wrong with the enforcement of the oil embargo.

But Thomson then went on to add that 'British oil companies have at no time been directly involved in the supply of oil to Rhodesia through Mozambique'.[9] This was misleading: Shell Mozambique had not *sold* directly to Rhodesia, it is true, since consignments for Rhodesia were purchased by Parry Leon & Hayhoe; but the oil had been *railed* directly to Rhodesia. Thomson's letter of 15 March 1968 therefore hardly clarified the situation – it apparently added further confusion to the whole affair.

On the following day the Commonwealth Secretary called on Wilson at Number Ten. In the wake of George Brown's resignation, however, it was hardly an opportune moment for a discussion of the intricacies of Rhodesian oil. The Prime Minister told Thomson that he had decided to amalgamate the Foreign Office and the Commonwealth Relations Office on the departure of George Brown. Thomson would therefore lose his post as Commonwealth Secretary, Wilson explained, but would be retained as a Cabinet Minister.[10] The question of oil sanctions faded into the background.

Within a few days of Thomson's meeting with the oil companies, the Rhodesian situation became dominated by the impending execution of three Africans in Salisbury. The

Queen, acting on advice from the British Government, issued a reprieve. But on 6 March 1968 the men were hung, and international anger against Smith reached new heights. Barbara Castle has claimed that the day after the executions Wilson told Cabinet that the 'only course was to intensify sanctions'.[11]

On 8 March the Overseas Policy and Defence Committee of the Cabinet discussed further action against Rhodesia. Wilson explained that the Government now had no option but to pursue a policy of 'activism' over sanctions, and various proposals to tighten the oil embargo, which had been discussed the previous year, were again raised. For some reason, however, Thomson seems to have relayed little of the information that he had gathered just two weeks earlier. He may perhaps have mentioned to the committee that 'suspicious' customers of Shell Mozambique had been diverting supplies to Rhodesia, but he does not appear to have suggested that Shell Mozambique had been contravening the Sanctions Order, nor that a swap arrangement had just been introduced.

Either Thomson had failed to inform Wilson of the facts, or else the Prime Minister had simply not appreciated the significance of what he was told. In either case it is clear that Wilson's overall view of the situation had not altered: although he was as concerned as ever about tightening the oil embargo, he seemed not to have appreciated the new developments which the Commonwealth Secretary had discovered in the previous few weeks. Barbara Castle has recounted that Wilson told Cabinet on 21 March that the Overseas Policy and Defence Committee was 'still studying the leak of oil to Rhodesia through Lourenço Marques,'[12] and wrote in her diary:

> The main sinner here is a French company and although Harold has tried to persuade de Gaulle, he has so far failed. I should have thought we could have conducted a vigorous campaign against France for this in every African country, but I doubt whether we are doing this.[13]

The Cabinet's Overseas Policy and Defence Committee met again to discuss sanctions on 28 March. 'We had forty minutes on Mozambique oil,' Crossman recalled, and Wilson

200

'is still convinced that we can strike a deadly blow'.[14]

What has not yet been revealed is the fact that this crucial meeting apparently considered a detailed 40-page paper on oil sanctions which had been prepared by the Foreign Office. Informed sources have suggested that the report *did* reveal that Shell Mozambique had been handling supplies for Rhodesia, and it would be surprising for a paper on this subject written just a month after the meeting with Shell and BP to omit this vital fact. The paper is also believed to have suggested that new arrangements had been made to remove Shell Mozambique from the supply chain. But at this point the Foreign Office had apparently not yet learned that this involved a swap arrangement with Total.

The Foreign Office report to the Overseas Policy and Defence Committee is clearly the key document which would resolve the 'who knew what' debate. If the report did indeed include the information which Thomson had been given by the oil companies at his meeting on 21 February 1968, then there would be no reason for Wilson, who chaired the Cabinet committee meeting, to claim that he was never informed that Shell Mozambique had been busting sanctions.

The Crossman Diaries note that during the Overseas Policy and Defence Committee meeting, the Prime Minister 'pleaded that we must now turn the heat on the French oil company which was putting oil into Rhodesia through Mozambique'.

Wilson's angry references to the role of the French oil company might appear to suggest that he *did* know about the Total swap. But closer examination suggests that this was probably not the case: the Prime Minister's references to Total were very similar to statements that he had made the previous year before the swap had ever been introduced. During early 1968 even Thomson himself appears not to have grasped the fact that a swap arrangement had been set up with Total, probably because this vital piece of information was deliberately withheld by Shell and BP at their meeting. Wilson's remarks on Total's involvement in sanctions-busting were therefore probably not prompted by any specific knowledge of the new swap arrangement, but originated from his mistaken belief that Total had all along been the major supplier of Rhodesia's oil. The Prime Minister's remarks about Total were therefore correct – but for the wrong reasons!

A bizarre interlude then occurred. George Thomas, Minister of State at the Commonwealth Office (not to be confused with George Thomson, the Commonwealth Secretary) publicly claimed that Total was supplying Algerian oil to Rhodesia – an accusation which obviously caused considerable unease at the Ministry of Power, where it was described as 'a mud-slinging exercise' (B/A 263). It is possible that Thomas had made the statement in the knowledge that Total had recently begun handling supplies for Shell and BP, but it is much more likely that he was simply basing his remarks on the assumption that the French-owned company had always been the major sanctions-buster.

Total's reaction was even more surprising than George Thomas's accusation. On 22 March 1968 the company's Paris headquarters issued a clear denial, stating that it was under a legal obligation not to supply oil to Rhodesia, either directly *or* indirectly, and that Total 'had in all circumstances conformed completely with the instructions which it had received from the Government on this matter'.[15] This was clearly untrue. Total's head office in Paris must have known that its South African subsidiary was supplying oil to Rhodesia through Parry Leon & Hayhoe, since quite apart from its own internal sources of information, the company had already received detailed reports from the Portuguese Government giving the exact quantities supplied by Total to Rhodesia. It is possible, however, that the head office of Total had not yet learned about the new swap arrangement with Shell and BP – which had only been introduced in Southern Africa just a few weeks earlier. For had Total's head office known about the arrangement it would surely have used this knowledge as ammunition to rebuff the British Government's 'mud-slinging exercise'.

Meanwhile further evidence on the identity of Rhodesia's supplier was provided in a new Portuguese Government report. Portugal, which was becoming increasingly angry at accusations that it was responsible for the failure of sanctions, again wanted to show that it was the international oil companies who were breaking the embargo. In March 1968 Portuguese Foreign Minister Dr Franco Nogueira asked Jardim to prepare a report on oil supplies to Rhodesia. Jardim collated figures showing the number of railway tank-cars leaving Lourenço Marques for Rhodesia: these had doubled from 5,500 in 1966 to 11,000 the following year. In 1967 the Southern African subsidiaries of Shell and BP were the largest suppliers (4,443 tank-cars), followed by Mobil (2,000), Caltex (1,697), Total (1,256), Sonarep (1,400), and South African Railways (240) (see facsimile document).[16]

The Portuguese report circulated remarkably quickly in Europe. Dr Nogueira sent it to the French Foreign Ministry, who immediately passed it on to Total, where it was studied by two executives responsible for Southern Africa (Duroc Danner and Etienne Dalemont). Total then apparently handed a copy to John Fearnley at the British Foreign Office and Dirk de Bruyne of Shell, who passed it on to BP. The British Government and the oil companies therefore had detailed evidence showing that Shell and BP had supplied over forty per cent of Rhodesia's oil during 1967. Since then, however, the swap arrangement had been introduced, and it was therefore felt that no further action was required from the British companies.

The cover-up was intensified. On 27 March 1968 Wilson told the House of Commons that sanctions were having 'a crippling effect' on Rhodesia; there had been evasions, he admitted, 'some of them based on unscrupulous, under-the-counter deals', but most members of the UN had respected sanctions.[17] One of the biggest difficulties had been oil, the Prime Minister admitted, and he went on to suggest, incorrectly, that the South African Loop route was still in use, and that oil was still being railed across from Lourenço Marques into South Africa, and then back into Mozambique for consignment to Rhodesia. 'The House will know of the transport complications,' he explained, 'where the railway crosses and recrosses the frontier.'[18] In fact, however, the Loop had not been used ever since the Direct Route had been introduced over a year earlier.

Later Thomson injected a good dose of hypocrisy into the debate. 'As I told the Commonwealth Sanctions Committee,' he explained to the House of Commons, 'no other country does more in the field of sanctions than the United Kingdom. I do not claim any special virtue for this. It is Britain's duty to do it . . . '[19] The Commonwealth Secretary added: 'If the general membership of the United Nations were to apply sanctions and supervise the activities of some of their business firms as well as Britain does, the impact of sanctions would be substantially and significantly increased.' Incorrect information and simple hypocrisy now seemed to have become the two staple ingredients of the Government's oil sanctions policy.

*　　　*　　　*　　　*

The Government had privately become resigned to the fact that nothing could be done to tighten the oil embargo. But after the executions in Salisbury in March, Britain needed to take some action against the rebel regime in order to fend off African demands for the use of military force. The time had come, Wilson decided, to extend the existing *selective* sanctions into *comprehensive* sanctions which would cover all of Rhodesia's foreign trade. These, Foreign Secretary Michael Stewart told Parliament, would 'impose on other countries the kind of obligations we ourselves are now fulfilling'.[20]

On 23 April 1968 Lord Caradon submitted the proposed resolution on comprehensive sanctions to the UN Security Council, again claiming that Britain's record was second to none. 'When there have been rare evasions by our nationals,' he explained, 'we have not hesitated to act'[21] – a remark that strikes a strange note coming from a Government which had just decided *not* to prosecute Shell Mozambique for a blatant breach of sanctions. Former Foreign Secretary Michael Stewart later justified Britain's decision to call for a widening of sanctions at this particular time: 'Since we could not tighten the oil embargo,' he admitted, 'it was important to do all we could over other sanctions.'[22] Nevertheless, had the other members of the UN known about the Government's discussions with Shell and BP, a major diplomatic row would undoubtedly have been precipitated.

The African members of the Security Council, as expected, pressed for tougher measures against the illegal Rhodesian regime. But on 29 May 1968 a slightly modified version of the British resolution was unanimously approved by the Security Council. The new resolution did not directly affect the oil embargo, since oil had been included in the earlier selective sanctions; but comprehensive sanctions did appear to indicate that the international community was united in applying tougher action against the rebel regime.

In Rhodesia, however, it was feared that the introduction of comprehensive sanctions might be accompanied by further moves to tighten the oil embargo. Genta, the Rhodesian Government procurement agency, therefore proposed that some of the paper-chases should be modified to disguise still further Rhodesia's oil supply routes. A letter from Richard van Niekerk (Mobil Rhodesia) to R. H. Maskew (Mobil South Africa) dated 2 September 1968 explained why the changes had to be made:

> You might consider that the procedure that we have adopted is unduly complicated and unnecessary, but as was conveyed to you when you were here, it is the wish of George's people [*a reference to George Atmore, Chairman of Genta*] that we involve and complicate this matter to a far greater degree than pertains at present in the hope that it will discourage an investigation.[23]

As the remainder of the letter went on to outline, modified paper-chases would now be used by Mobil to supply petrol, diesel, and lubricants, although the basic pattern – and in particular the use of Parry Leon & Hayhoe, and Genta – would remain the same.

Discussion of comprehensive sanctions had led to a further meeting between the British Government and the oil companies on 15 May 1968, at which a Ministry of Power official was given full details of the new swap arrangement by Shell and BP. Alan Gregory began by asking Shell and BP to confirm 'the general impression' that he had gained that Total was 'a prominent factor' in the new arrangements taking Shell Mozambique out of the supply chain (B/A 264). John Francis agreed. Gregory then went on to seek confirmation that 'Shell and Total had, in effect, exchanged a bit of the market outlet in Mozambique'. Francis again agreed in general terms, and asked whether Shell should make more specific inquiries on this point. Gregory, according to Shell's record of the meeting, then 'said that in general he recognised that we [i.e. Shell] would not wish to be too involved with the details, and nor, for that matter, would HMG'. This was a startling remark: the Ministry of

Power appeared not only to be discouraging Shell from finding out how its South African subsidiary was supplying Rhodesia, but also to be suggesting that the British Government had even less interest in the subject; indeed, he was almost admitting that the Government would prefer *not* to know how the Total swap operated.

Even the information which had been obtained, however, did not seem destined to get very far within Whitehall. For at one point in the meeting Gregory asked Francis whether the Ministry of Power could give certain information 'on a purely personal basis' to Robert Clinton-Thomas at the Commonwealth Office, so that 'if there were any further signs of Ministers wishing to sound off on this subject, the appropriate discouraging noises could be made' (B/A 264). The fact that Clinton-Thomas was provided with material on a 'personal' basis suggests that it was not intended for Thomson's eyes – an impression which appears to be confirmed by the proposal that the Minister should be fended off with nothing more than 'discouraging noises'.

Gregory also recounted that on 22 April a French Foreign Ministry official (Jourdan) had called on the British Embassy in Paris, and had 'provided an explanation of the Genta purchasing machinery' (B/A 263). The French civil servant had said it would 'be appreciated if the British Press (and perhaps politicians also) would stop accusing the French of being the main cause of the leakage' from Lourenço Marques to Rhodesia. Gregory then told the Shell and BP representatives that the British Government 'must avoid any accusations against the French which would cause the French Government to investigate more closely the swap arrangement' (B/A 264). For if Britain continued to allege that Total was the major sanctions-buster, it was feared that the French Government would publicly reveal the existence of the swap arrangement between Total and the British-owned companies in South Africa. The introduction of the swap, and the subsequent decision to end pressure on the French Government therefore made it even less likely that there could now be an international initiative to tighten the oil embargo.

Within one month the Prime Minister appeared to be following Gregory's advice against making accusations against Total. Ben Whitaker, MP, had asked whether the Government would be willing to 'tell those individual companies, such as the Compagnie des Pétroles Française [Total], who are torpedoeing sanctions, that they must choose between trading with us or Mr Smith?' Wilson, in a dramatic reversal of his earlier line, replied: 'I do not think there is any ground for complaint against the French Government in this matter.'[24]

Meanwhile Britain continued its efforts to reach a political settlement to the Rhodesian problem. Early in October 1968, Richard Crossman recalled, Wilson gave his Cabinet 'a good long lecture on how Smith needed a solution much more than we did, how economic sanctions had reached their maximum, and how the South Africans wanted a settlement.'[25] Crossman added: 'It's my impression that this lecture didn't convince a single member of Cabinet.'

On 8 October 1968 Wilson flew to Gibraltar for talks with Smith on another Royal Navy vessel, this time the *Fearless*. The talks, like the *Tiger* negotiations nearly two years earlier, were unproductive, and Wilson's proposals, which in any case would not have led to majority rule for many years, were finally rejected by Smith. On 22 October 1968 Joan Lestor MP asked Wilson whether there would be any relaxation of sanctions. The Prime Minister confirmed that there would be no relaxation: sanctions were to continue.[26]

*　　*　　*　　*

A fascinating development took place a few days after the failure of the *Fearless* talks. The South African Minister of Economic Affairs suddenly sent a 'secret' letter to the South African subsidiaries of Shell and BP, reaffirming 'the verbal directive given to the oil companies . . . that the Government requires the oil companies to refrain at all times from imposing . . . conditions or reservations of whatever nature in respect of the use, resale or further distribution of' oil products (B3.3). This was a reference to the 'conditional selling' legislation, which the oil companies argued made it impossible to cut off sales to Rhodesia. It will be recalled that many years later Mobil used the existence of this law to counter *The Oil Conspiracy* allegations.

The significance of this letter from the Ministry of Economic Affairs is as follows: the South African 'conditional selling' legislation can only come into effect *after* a formal written notice has been issued, *a step which had not been taken before this letter was sent*. This meant that in the months after UDI, when oil from the South African subsidiaries of the oil companies began to flow into Rhodesia, there was nothing to prevent the oil companies from refusing to sell to customers whom they believed to be purchasing supplies for Rhodesia. Yet the fact was that it was not until October 1968, nearly three years after UDI, that written notice was given to the oil companies – a clear sign that oil companies may have been deliberately using the existence of South African legislation to justify the continued flow of their oil to Rhodesia. Certainly, as Bingham points out, 'the South African law on conditional selling was no doubt welcomed by the companies' South African employees who were sympathetic to the Rhodesian regime' (B3.5).

Further evidence to support this view had emerged earlier in 1968. At the meeting with Thomson on 21 February 1968, the oil companies told the Commonwealth Secretary that their subsidiaries in South Africa were obliged 'to sell to any South African buyers without conditions' (B/A 256). Privately however, they seem to have been less convinced of the force of this argument, for in a note of a discussion between Shell and BP executives on 16 February, just five days before the meeting, it was 'agreed that the South African instruction [on conditional selling] that we seemed to be *sheltering under* and *using as an excuse* was only verbal, had been given by Kotzenburg [the Secretary for Commerce] *and could well be repudiated*' (B6.65). (The emphasis is added.)

The fact that later in the same year the South African Government did issue the oil companies with written notice at least raises the possibility that it was prompted to do so informally by the oil companies themselves. Quite clearly, once this written notice had been sent, the oil companies were in a stronger position to continue supplying Rhodesia: they could now argue, as Mobil did in 1976, that 'conditional selling' laws *forced* them to sell to customers who sold to Rhodesia.

Although 'conditional selling' legislation helped to shield the oil companies when they came under public attack in

1976, its importance may well have been exaggerated. Quite apart from the fact that for the three years before written notice was given, the oil companies could have refused to sell to its South African customers with little fear of immediate to their South African customers with little fear of immediate Government action, the law did not apply to 'anyone who is obviously acting as a mere agent for a Rhodesian purchaser' – as Mobil's lawyers were later to admit.[27] At their meeting with Thomson on 21 February 1968, the oil companies had insisted that they were obliged to sell 'without condition'; but this questionable argument also failed to explain why Shell Mozambique (which, of course, was not subject to South African law) was forced to handle the purchases – and thus breach the Sanctions Order. There was therefore strong evidence to suggest that the oil companies were indeed 'sheltering' under rather dubious interpretations of South African legislation.

* * * *

In January 1969 the Portuguese Government again produced evidence showing that it was the British and American companies who were supplying most of Rhodesia's oil. Figures provided for the month of November 1968 suggested that the Southern African subsidiaries of Shell and BP had sent 517 railway tank-cars to Rhodesia, with Caltex (293), Mobil (217), and Total (68) sending the remaining traffic (B/A 268). Again the British Government discounted this report. But it was nevertheless decided that a further meeting should be held with the oil companies and on 6 February 1969 Frank McFadzean (Shell) and Robert Dummett (BP) were summoned to Whitehall.

Thomson was no longer Commonwealth Secretary, because of the recent merger of the Foreign and Commonwealth offices, and he was now Minister without Portfolio. But in the absence of the Foreign Secretary, he was asked to lead the Government side of the discussions. Thomson, according to the Government record of the meeting, began by saying that 'it would be useful to know how these [swap] arrangements were working' (B/A 268). He had been told very little about the swap at his previous meeting with the oil companies, almost exactly a year before, but his opening remark suggests that he had learned more about the situation since then.

McFadzean, the Shell Managing Director, explained exactly how the swap worked: 'When an order was placed with Shell/BP South Africa by the South African broking firm of Parry Leon & Hayhoe on behalf of the Rhodesian oil-purchasing organisation Genta, the supply was actually made by Total's associate company in Lourenço Marques' (B/A 269). Alan Gregory, of the Ministry of Power, later confirmed to Bingham that 'the supply arrangement (and the fact that the suspect buyers remained *customers* of Shell South Africa) was explained in a much clearer, and indeed rather different, way' in February 1969 than it had been at his own meeting with John Francis (Shell) and Arthur Sandford (BP) in May 1968 (B6.84).

Even after Thomson had learned exactly how the Total swap operated, however, the Minister still felt that it was consistent with Government policy and the sanctions legislation. Thomson told Bingham that he was 'under no illusion' that he was reducing the flow of oil into Rhodesia (B6.76). It was 'a second best solution.'[28] But it did mean that 'the Government was ensuring that British companies under their jurisdiction were observing British law – and that was of capital political importance in maintaining the pressure on Smith for a negotiated settlement'.[29]

Thomson argued that Britain was not actually able to stop oil reaching Rhodesia. 'What we did have the power to do,' he later explained, 'was to bring the maximum diplomatic pressure to bear on the French Government to make them stop the flow of French oil into Rhodesia.'[30] In fact, as the Ministry of Power had realised, the introduction of the swap arrangement really made it even *more* difficult to put pressure on the French Government to tighten sanctions. Thomson's position also lacked a certain logic. The South African subsidiaries of Shell and BP, he had argued, were beyond the control of the British Government. But exactly the same argument could be applied to Total South Africa. If the British Government had been unable to stop Shell South Africa from *initiating* the swap, then there was little reason why the French Government should have been able to prevent Total South Africa from *implementing* the arrangement.

The swap was accepted by the Government; but to the oil companies it was simply a 'cosmetic' device which 'appeared to be merely an alternative means of supply'[31] (B14.19). And the fact that the Total swap did not actually reduce supplies to Rhodesia, Bingham explained, 'induced among some of those most directly concerned (notably Mr Francis and Mr Walker) a belief that compliance with the Sanctions Order was to be regarded as a matter of form rather than substance, that it was the letter which mattered, not the spirit' (B14.18).

Thomson, however, saw the swap in rather different terms. For him it meant that the Government could now say that no 'British' oil was reaching Rhodesia. His decision to approve the swap was therefore in line with his stated view that 'a great deal of politics normally is a problem of *presentation*'. (B6.76). It was a surprising decision nonetheless from a politician who had earned a reputation for honesty. For only a year before the swap was condoned, Thomson had written to the Labour Party to point out that 'new sanctions measures which look *presentationally* good outside Rhodesia when approved, but which then proved ineffective in practice inside Rhodesia, reduce the international credibility of sanctions and serve to increase the White Rhodesians' confidence in their capacity to defy the world'[32] (emphasis added).

*　　*　　*　　*

Thomson's meeting of 6 February 1969, at which full details of the swap were given, again raises the 'who knew what' question. The official record of the discussion was sent to Michael Palliser at Number Ten. Wilson, however, claims that there is no record of either his Private Secretary or himself having seen it. 'That may sound bizarre,' Wilson told Parliament, 'but hundreds of documents, telegrams, despatches, notes of meetings, reports and assessments from the Foreign Office come in every week.' [33] This particular record, he complained, 'was not marked urgent or highlighted in any way'. Palliser had just been appointed Minister at the British Embassy in Paris, and so was frequently away on visits to the French capital, and he was also heavily engaged in preparing for the Prime Minister's coming visit to Nigeria and Ethiopia. Wilson therefore excused Pal-

liser and argued that 'there is no reflection on him whatsoever' for failing to recognise the importance of the document.[34]

Thomson called at Number Ten shortly after the February
1969 meeting. But at this time both politicians were more
concerned about a crisis that had developed over Britain's
relations with the European Community, and there was
little if any discussion of Rhodesian oil.[35] In any case, it
seems that nothing about the Total swap appeared in the
Prime Minister's Rhodesian files. Wilson later claimed to
have gone through these documents – 'four bulky files, in all
about 8 or 9 inches thick' – and to have found nothing on
the swap arrangements.[36]

Wilson has recently argued that Thomson never really
understood the implications of the Total Swap. The official
minute of the meeting, Wilson claimed, 'suggests that there
was little realisation of the import of the disclosure.'[37]
Certainly the Government record, which is written in the
very low-key style beloved of civil servants, hardly gives the
impression that the officials present understood the full
significance of what they were being told; it also seems
that their questioning of the oil companies was very perfunc-
tory.

Wilson has also claimed that neither the Foreign Secretary
nor any Foreign Office civil servants realised the importance
of Thomson's meeting. If Foreign Secretary Michael Stewart
had read the report of the February 1969 meeting he 'would
have dropped everything and come round to see me, perhaps stopping for a moment to telephone me to say that he
was on his way.[38] Stewart would no doubt have also 'put it as
the first item on his weekly general report to Cabinet on
Foreign Affairs'.

Stewart himself, however, later disputed Wilson's explanation. He recently told the House of Commons that he
was aware of the swap, at least by February 1969, and had
felt that it was a justifiable arrangement. He had admitted,
however, that he never raised the Total swap with Wilson
when he was Foreign Secretary.[39]

It is clear that the existence of the swap was known to a
number of senior Foreign Office officials in London, since
the minutes of the February 1969 meeting were sent to three
civil servants who attended (Sir Leslie Monson, James

212

Bottomley, and Robert Clinton-Thomas). In addition the minutes also went to the Private Secretary of Denis Green-hill, then permanent head of the Foreign Office, and later to become Lord Greenhill and a Director of BP. Maurice Foley, then Parliamentary Under-Secretary at the Foreign Office, had also been present at the discussions and he too received a record of the meeting.

It is disturbing, however, that news of the swap was apparently not received by one of the Foreign Office Ministers most deeply involved in dealing with Britain's sanctions policy. Lord Caradon, who as Britain's Representative at the UN had played a key role in obtaining international support for comprehensive sanctions in 1968, claims that he was not informed about Thomson's discussions with the oil companies. When the facts publicly emerged a decade later he described it as 'a disgraceful story'.[40] He has also suggested that he would have actually resigned from the Government if he had known about Thomson's acceptance of the swap.[41] The fact that Lord Caradon's strong views on Rhodesia were well known, and that he could have been expected to have objected strongly to the swap, makes it possible that a decision was made in London *not* to inform him about these developments – or else to 'hide' the information in a long document.

Neither Thomson nor Wilson emerges unscathed from the affair. Thomson seems to have been manipulated by both the oil company executives and the civil servants at the Ministry of Power: the news that Shell Mozambique had been busting sanctions, and that a swap had then been set up with Total, appears to have been very deliberately broken to him step by step, each revelation being carefully presented in a way that made it more difficult for him to lodge objections. At the same time Thomson seems to have made little effort to probe the matter, and the records available suggest that he was insufficiently aware of the great political sensitivity of the oil embargo. Shell and BP therefore succeeded in putting 'the Government's fingerprints' on the swap.

Wilson comes out equally badly. If, as he claims, he did not know about the swap, then much of the responsibility must be his own. Rhodesia was the most sensitive foreign policy issue that Wilson had to face in office; sanctions were the key element to force an end to the rebellion; and the oil embargo was the most important form of sanctions. The Prime Minister had a responsibility to ask why the oil embargo was failing. Winston Churchill's comment over Britain's lack of preparation for Singapore's defence during the War sums it up: 'I did not know; I was not told; I should have *asked*.'

Wilson claims that the Total swap was never discussed by Cabinet. Neither Thomson nor Stewart, the two Cabinet Ministers who have admitted to knowing about the arrangement, seem to have raised it, despite the fact that there were a number of discussions on sanctions. A copy of the record of the 6 February 1969 meeting was sent to the Private Secretary to the Cabinet Secretary, Gruffydd Jones, but again, Wilson has claimed 'it was not marked for special attention by him'.[42] 'If the Cabinet Secretary had realised its importance,' Wilson added, 'I am sure that he would have come steaming in right away to insist that it go to the Cabinet.' Wilson also pointed out that the Attorney General, Sir Elwyn Jones, did not know about the meeting. Sir Elwyn Jones, who was among the most militant Cabinet Ministers over sanctions, would no doubt have felt that his duty was 'clear', Wilson added, had he known about the Total swap.[13]

On 7 March 1969, just a month after Thomson's meeting, the Cabinet had a long discussion on Rhodesia, but it seems that the oil embargo was not discussed. One month later, when the Cabinet's Rhodesia Committee met, Wilson asked for a full report on sanctions from the Foreign Secretary. Stewart duly presented a detailed report, nearly thirty pages long, but apparently this contained not a single reference to either the February 1969 meeting or the Total swap arrangement. It is fascinating to speculate what would have happened if the swap arrangement had been discussed by the Cabinet in 1969. Wilson himself has claimed that 'the future conduct of every Ministerial meeting involving Rhodesia would have been entirely different'.[44] Certainly it would have lead to strong pressure from militant members of the Cabinet for further action to tighten sanctions. Quite possibly the collective decision of the Cabinet would still have been to condone the swap. But at least this would have meant that Britain's Rhodesian policy could have been discussed on a more realistic basis.

NOTES

1 *Guardian*, 29 August 1978.
2 Thomson's statement of 6 September 1978.
3 BP's Submission to the Bingham Inquiry, 27 September 1977, p. 72.
4 *Guardian*, 26 June 1978.
5 Hansard, Lords, 9 November 1978, col. 460.
6 Wilson's statement of 7 September 1978.
7 Hansard, Commons, 7 November 1978, col. 742.
8 Wilson's statement of 7 September 1978.
9 Wilson's statement of 7 September 1978.
10 Wilson's statement of 7 September 1978.
11 *Sunday Times*, 10 September 1978.
12 *Sunday Times*, 10 September 1978.
13 *Sunday Times*, 10 September 1978.
14 Richard Crossman, *Diaries of a Cabinet Minister*, vol. ii, p. 744.
15 Statement by Total, 22 March 1968.
16 Letter from Arthur Sandford (BP) to Robert Dummett (BP), 11 April 1968 (see illustration 2).
17 Hansard, Commons, 27 March 1968, col. 1565.
18 Hansard, Commons, 27 March 1968, col. 1566.
19 Hansard, Commons, 27 March 1968, col. 1688.
20 Hansard, Commons, 24 April 1968, col. 235.
21 Security Council debate, 23 April 1968.
22 Interview, 10 April 1979.
23 *The Oil Conspiracy*, p. 36.
24 Hansard, Commons, 27 June 1968, col. 810.
25 Richard Crossman, *Diaries of a Cabinet Minister*, vol. iii, p. 217.
26 Hansard, Commons, 22 October 1968, col. 270.
27 Report by Edward Fowler (General Counsel of Mobil), 10 September 1976, p. 56.
28 Granada Television, *World in Action*, 7 August 1978.
29 Thomson's statement of 6 September 1978.
30 Thomson's statement of 6 September 1978.
31 Letter from Sir Frank McFadzean to *The Times*, 5 September 1978.
32 Letter from Thomson to Labour Party International Committee, 29 December 1967.
33–38 Hansard, Commons, 7 November 1978, cols. 744–5.
39 Interview, 10 April 1979.
40 Interview, 11 August 1978.
41 Granada Television, *World in Action*, 7 August 1978.
42 Hansard, Commons, 7 November 1978, col. 745.
43 Hansard, Commons, 7 November 1978, col. 746.
44 Hansard, Commons, 7 November 1978, col. 746.

11   Business As Usual

February 1969–February 1976

Shell and BP executives emerged from their meeting with Thomson on 6 February 1969 with a new sense of confidence. The British Government had condoned the swap arrangement with Total, admitting that it was 'legally sound', and the oil companies could therefore continue to supply their Rhodesian subsidiaries. Exactly one week after this crucial meeting, Shell took an important initiative over a major new investment project inside Rhodesia.

On 13 February 1969 Hugh Feetham, a Shell executive dealing with Southern Africa, wrote to Shell South Africa, pointing out that plans for a new lubricant blending plant at Salisbury had now reached a stage at which Shell Rhodesia needed 'Central Offices' advice' (B12.24). But to avoid Shell Centre having actually to send out a technician to Salisbury, Feetham suggested the rather transparent expedient of sending the expert (Moscisker) to South Africa instead, and then arranging for Shell Rhodesia to send someone down to Johannesburg 'in order to obtain the benefit of his know-how'.

The confidence of the British oil companies was strengthened still further by the Conservative victory in the June 1970 election. The Tories, after all, had even less enthusiasm than Labour for Rhodesian sanctions. As Bingham diplomatically put it: during the early 1970s 'the enforcement and monitoring of sanctions came to assume a lower Governmental priority than they had previously done' (B6.87).

Right from UDI the Conservatives had been unenthusiastic about sanctions, many Tories having abstained or even voted against the oil embargo. But once sanctions had been established, and approved by the UN Security Council, it

216

would have been extremely difficult for Britain to have lifted them unilaterally in the absence of an internationally acceptable settlement to the Rhodesian problem. Sanctions were therefore retained by the incoming Conservative Government in 1970, but even less effort was made to monitor their effectiveness. Sir Alec Douglas-Home, the Foreign Secretary, has revealed that during his period in office 'we never discussed in Cabinet committee, or in Cabinet, the question of sanctions and oil'.[1]

The new Government appears to have known nothing about Thomson's discussions with the oil companies and the introduction of the Total swap arrangement. Former Prime Minister Edward Heath told the House of Commons: 'We did not see the previous Government's papers and we were therefore not told about the swap.'[2] Sir Alec Douglas-Home has admitted that he personally 'never knew of any British involvement' in oil supplies to Rhodesia,[3] and this claim was backed by Joseph Godber, a Minister of State at the Foreign Office, who said that the existence of the swap 'was never disclosed' to him.[4]

The civil servants, however, have disputed their masters' ignorance. Sir Denis Greenhill, Permanent Head of the Foreign Office from 1969 to 1973, received minutes of the 6 February 1969 meeting, and has admitted that he knew about the swap.[5] It is therefore strange if this information was not passed on to the incoming Tory Foreign Secretary. Greenhills deputy, Martin Le Quesne, who had attended the same meeting, has also suggested that incoming Ministers would indeed have been fully briefed on all aspects of the Rhodesian situation.[6]

Rhodesia was the most important foreign policy issue faced by the new Conservative Government. Indeed, soon after Sir Alec Douglas-Home took over as Foreign Secretary, a thorough report was prepared for him on the prospects of a Rhodesian settlement, including a detailed section on sanctions. It is not known whether the oil embargo was discussed in the document; but if details of the swap *were* included, then there is no excuse for Conservative Ministers to claim ignorance of it. If, on the other hand, the facts about the swap arrangement were *not* given, then the civil servants involved were guilty of withholding vitally important information from their Ministers.

By this time the Ministry of Power had been absorbed into the Ministry of Technology, although Angus Beckett continued to head its Petroleum Department. Beckett himself has said that he 'just cannot believe that Conservative Ministers weren't briefed about the swap'.[7] Again, however, the politicians claim to have been ignorant about the oil embargo: Geoffrey Rippon, Minister of Technology for the first five weeks of the Conservative Government, has said that he was not told about the swap;[8] John Davies, who succeeded Rippon, has claimed that he was 'totally uninformed about the whole thing'[9] (– this in spite of the fact that he was particularly well placed to have understood the background to the oil embargo, through having been Managing Director of Shell Mex and BP until 1965); and Sir John Eden, Minister of State to both Rippon and Davies, has also said that he was never told about the swap when he was in office.[10]

Even after the introduction of the swap arrangement, however, it seems that Shell Mozambique did not completely sever its Rhodesian connection. Jardim has claimed that Total sometimes used the Shell Mozambique depot at Lourenço Marques to store oil destined for Rhodesia. The provision of this storage facility (known as 'hospitality' in the trade) may well have made Shell a partner in sanctions-busting.[11] There is also other evidence to suggest that Shell Mozambique continued to handle supplies for Rhodesia even after the swap had been set up. On 5 November 1970, for instance, Freight Services wrote to Shell Mozambique to inform the company that the keys to the Rhodesia Railways tank-cars were being changed.[12] Presumably Shell Mozambique was therefore still handling some oil for Rhodesia. Bingham has concluded that during the period of the Total swap, deliveries to Rhodesia from Shell Mozambique 'were at a very much reduced rate' (B8.5): but as long as Shell Mozambique was sending *any* oil to Rhodesia, it was still contravening the Sanctions Order.

Meanwhile, the Conservatives made yet another attempt to negotiate with the Smith regime. In November 1970, just a few months after the Tories took over, Sir Alec Douglas-Home told Parliament that secret talks had begun in Pretoria between British and Rhodesian diplomats. Further negotiations culminated in the agreement signed on 24

218

November 1971, which, although it would not have led to majority rule for many years and was widely regarded in Africa as a sell-out, was intended to lead to a return to legality. Under the terms of the accord, a commission of inquiry was to determine whether the proposed settlement had the backing of the Rhodesian people; this sparked off the establishment of Bishop Abel Muzorewa's African National Council, which campaigned for a rejection of the plan. The Pearce Commission subsequently reported in May 1972 that the settlement proposals had little support among the African population, and the plan was scrapped.

Sanctions still remained in force, but the Conservatives had little enthusiasm for making them bite. Julian Amery, a Minister of State at the Foreign Office has admitted that the climate of opinion in the Conservative Party was 'rather relaxed' over sanctions.[13] He explained:

> I would reject completely any idea that there was connivance with law breaking and anything of that kind. On the other hand, I don't think we would have been witch-hunting in a matter of that kind. It would have been the distinction between tax avoidance and tax evasion.[14]

* * * *

Joshua Nkomo, the veteran nationalist leader, had been detained for more than a decade by the Smith regime. For most of the time he languished at the remote Gonukadzingwa camp, not far from the Mozambique border. From the camp, which was close to the railway line for Lourenço Marques, it was possible to see the trains of oil-cars being hauled into Rhodesia. It was a bitter reminder to the detained nationalist. For just before UDI, Wilson had visited Rhodesia, met Nkomo, and told him not to worry; 'everything is under control', the British Prime Minister had apparently said; UDI would never come – because Wilson had 'threatened Smith with an oil embargo'.[15] The daily trains bringing oil into the beleaguered country must therefore have been a particular vivid symbol to Nkomo of Britain's failure to deal with the Rhodesian rebellion.

The oil continued to flow. Rhodesia's imports rose steadily from 7,700 barrels a day in 1968 to 10,900 barrels a day five years later. Rationing, which had been introduced shortly after the embargo, was little more than a minor inconvenience now. Rhodesians gleefully pointed out that petrol was actually cheaper in Salisbury than in London, and that at a slightly higher price motorists were able to buy as much unrationed petrol as they wanted. Rationing itself was eventually abolished on 12 May 1971. For the oil companies it was 'business as usual'.

* * * *

That nothing had fundamentally changed since UDI is illustrated by the fact that Shell's Southern African subsidiaries apparently acted as *sole* supplier of oil products to the Rhodesian armed forces. Here was an army and air force in open rebellion against the Crown; yet it was Shell, a British-owned company, which continued to provide the fuel which was so essential to give this armed force its mobility.

The Bingham Report also reveals that the London head office of Shell knew about these sales. For on one occasion, in the autumn of 1972, Shell Centre was asked – 'on the grounds that the Rhodesian Armed Forces were 100 per cent customers of Shell Rhodesia' – to help organise the 'transport requirements' of General Keith Coster, the retiring Commander of the Rhodesian army, who was then on a secret visit to London (B12.7).

Shell has argued that its Rhodesian subsidiary had been beyond its control since UDI. Yet the fact that Shell Rhodesia asked Shell Centre in London to help the retiring Army Commander – and stated the reason for the request in such embarrassingly explicit terms – suggests that Shell Rhodesia hardly expected any raised eyebrows at head office. The incident certainly suggests that Shell Centre had always known about these sales, and had never registered any unease over the fact that the supplies its Southern African subsidiaries were making were ending up by fuelling a rebel army.

The South African subsidiaries of the oil companies continued to use the paper-chase to supply Rhodesia. When Parry Leon & Hayhoe was absorbed into Freight Services in 1969, the new company took over as middleman in this lucrative trade; but otherwise the basic supply procedure remained unchanged. By this time a whole new vocabulary had been developed to deal with this clandestine trade. Confidential oil company papers would talk coyly of *'Our Friends to the North'*; more formal documents referred to *'Extraneous Demand'*; the word *'Freight Services'* or simply *'FS'* would crop up as the name of a country; in conversation the expression *'Grey Area'* was used for Rhodesian trade; and the London offices of the oil companies became accustomed to sending shipments of refined products to Lourenço Marques with the shipping papers marked *'Final destination to be determined'*. But behind these rather clumsy attempts at disguise, oil supplies continued unabated.

The established marketing companies in Rhodesia were extremely jealous of competitors coming into the market, or of existing marketers grabbing a larger share. Immediately after UDI it was therefore decided that each of the companies should be permitted to retain the same share of the market that they held at the time the oil embargo was introduced – an arrangement that was policed by Genta. But there were attempted infringements: in June 1966 Caltex South Africa unsuccessfully tried to take over the entire Rhodesian market by offering supplies at a discount; and many years later a BP report revealed that 'in 1970 Esso offered to supply one hundred per cent of Genta volumes at heavily discounted prices, and were turned down in favour of a continuing relationship with the established marketers' (B8.39). During the early 1970s Sonarep also formed a Rhodesian subsidiary, the Rhodesia Oil Company, but this never actually began marketing. Manuel Boullosa, the owner of Sonarep, has revealed that in 1973 he had plans to triple the output of his refinery to 50,000 barrels a day in order to supply Rhodesia's entire requirements – but opposition from the established oil companies forced Boullosa to drop his ambitious plan.[16]

The South African subsidiaries of two other oil companies are believed to have supplied lubricants to Rhodesia: Esso South Africa apparently used Virginia Oil and Chemical Company as an intermediary. In 1971 Virginia claimed that 'Esso used Virginia Oil and Chemical to export its products to Rhodesia. But the products were sent directly by Esso to Rhodesia and not via Virginia's depot.'[17] Castrol lubricants have also continued to be marketed in Rhodesia since UDI.

Bingham reported that he had taken up the matter with the head office of Castrol in Britain, 'but without obtaining any clear picture as to how these products reached Rhodesia' (B p.v). Subsequently Foreign Secretary Dr Owen told Parliament that the Director of Public Prosecutions was studying the question of whether Castrol or its parent, Burmah Oil, might have contravened the Sanctions Order. The fact that Dennis Thatcher, husband of the Prime Minister, was a Director of Castrol from 1967 to 1976 has given the matter a particular political sensitivity.

* * * *

The established oil companies did well financially after UDI. Robert Barrie, the General Manager of Shell Rhodesia, wrote to John Francis at Shell Centre to report that the company had 'a bigger slice of the cake through present methods' than it would have had if the Umtali refinery had come back on stream' (B8.66). In other words, Shell had actually benefited from the developments that followed UDI. Barrie went on to explain one of the reasons for Shell's success: 'We have created, I think it is fair to say, an outstanding rapport with Genta, the Government fuel procurement agency'. BP seemed equally pleased with the performance of its Rhodesian interests: a BP report prepared in February 1974 commented that its subsidiary in Rhodesia was 'a very profitable enterprise' (B8.39).

Bingham provides highly sensitive statistical data showing that Shell and BP Rhodesia made total profits of R$17.4 million (about £13 million) in the period from 1965 to 1975 (B12.20). Mobil did even better. Confidential Mobil Rhodesia documents, reproduced in *The Oil Conspiracy*, show that its profits from 1965 to 1973 amounted to R$7.6 million (about £6 million).[18] The Mobil figures also reveal that its financial situation improved considerably after UDI: profits per barrel more than tripled. The Rhodesian Government had presumably helped to ensure the full cooperation of the oil companies by enabling the marketing firms to obtain good profits on sales.

Ironically, the difficulty for the oil companies was not in importing the oil supplies but in repatriating their Rhodesian earnings. Under Rhodesian regulations the remittance of

222

profits to Britain and the United States was prohibited after UDI, in retaliation against sanctions imposed against the Smith regime. The problems – and a number of possible solutions – are very frankly discussed in a highly revealing memorandum prepared by Cecil Low, Financial Director of BP South Africa, for Denys Milne, the British Managing Director of BP Southern Oil. Fortunately a copy of this crucial report was included in the Sandford File.

Low's paper, which is dated 6 May 1974, provides a fascinating insight into the ways in which multinational companies are able to utilise their international links. The memorandum discusses three different proposals to remit the Rhodesian profits of Shell and BP to head office. First, it points out that remittances are permitted to some countries – and singles out Holland, where Royal Dutch Petroleum is based, for special mention: 'Naturally the Rhodesian management were aware that if the Rhodesian operation could be transferred to South African or, for that matter, Dutch ownership, it would be possible to circumvent the restriction on the payment of dividends.'[19] But Low adds that such a change would obviously have to have the approval of Shell Centre in London.

Secondly, Low suggests that the Rhodesian authorities might be willing to make a special arrangement to permit the transfer of profits to London in return for 'assurances on supply arrangements'. This, he adds, 'could be in Rhodesia's national interest'. Shell and BP, in other words, would receive their Rhodesian profits on condition that they guaranteed to continue oil supplies to the rebel regime.

A third possibility was 'the loading of transfer prices in such a way as to reduce Rhodesian profitability and increase profitability in South African hands'. These additional South African profits could then be forwarded to London. A variation of this proposal which had been 'adopted by the American companies' was 'to centralise as much of their Rhodesian administration in South Africa'. Low added that at one stage Caltex South Africa was charging Caltex Rhodesia R\$400,000 a year (about £300,000) for this 'service'.

Low's memorandum concludes with a comment which reveals the true attitude of many executives of the South African-based oil companies. The Smith regime, he admitted, had treated the oil companies well: 'A black

223

Government finding themselves in such circumstances and in a buyers' market would have taken a harsh line, whereas the Rhodesian oil industry has been treated, in my view, very generously.' This is blatant racism – implying, as it does, that blacks are somehow more likely than whites to exploit a situation for financial gain.

It is not known whether Shell and BP managed to repatriate their Rhodesian profits by the routes suggested by Low. But some of the profits that had been built up since UDI were used on the construction of the R\$550,000 (£400,000) lubricant blending plant at Willowvale, on the outskirts of Salisbury. When the plant was built in the early 1970s, after the Conservatives had come to power, the head offices of the oil companies had become more brazen about their dealings with Rhodesia. In March 1972, for instance, John Francis of Shell Centre personally visited Rhodesia – presumably feeling able to dispense with the caution which three years earlier had led the company to send the technician who advised on the project to Johannesburg instead. On his return to London, Francis wrote a confidential aide-mémoire indicating that he would prefer to drop the word 'Shell' from the name of the company being formed to operate the plant, since 'the publicity given outside Rhodesia to the Shell/BP investment in the Luboil plant would be minimised, thus saving the shareholders embarrassment elsewhere' (B12.26). It was therefore proposed that the firm should shelter under the innocuous name of Rhodesian Oil Products. In July 1974, at the time when the plant was opening, three British BP officials visited Salisbury and formed the impression that the project was 'somewhat gold-plated', because it absorbed capital which would otherwise have been earning little interest (B12.22).

Total was the only one of the international oil companies which had little difficulty in repatriating its Rhodesian profits, forwarding them regularly from Salisbury to Paris, where they were incorporated in the accounts of the Compagnie Française des Pétroles. Hidden away in CFP's 1977 Annual Report, however, is a small note explaining that the accounts of Total Rhodesia were no longer to be incorporated in the results of the group. This change took place just after widespread allegations of sanctions-busting were being made against the oil companies, and Total's head office may therefore have realised that it was perhaps wiser *not* to incorporate its Rhodesian profits into the balance sheet.

Trade with Rhodesia, however, also generated profits inside Rhodesia, since the South African subsidiaries of the oil companies profited from the fees obtained for refining the crude oil that was ultimately exported to Rhodesia. A Shell document estimated that this earned Shell and BP R\$3 million (1,950,000) in 1975. The two British-owned oil companies also made R\$1million (£660,000) profit from handling changes on trade with Freight Services (B11.8). Finally, the international oil companies gained a profit on the crude oil which they shipped to South Africa.

*   *   *   *

Before the publication of the Bingham Report it was generally assumed that there had been little contact between the head offices of the Rhodesian oil companies and their Rhodesian subsidiaries since UDI. But it is now clear that this was not the case. John Francis visited Rhodesia on at least three occasions (November 1967, March 1972, and March 1975); in May 1974, Turpin, an official from Shell Centre, carried out an inspection of facilities at Rhodesian airfields; and two months later, a team of three British BP executives (Denys Milne, Alan Robertson, and Neil Webber) arrived in Salisbury (B12.8). During these visits, Bingham has revealed, information was often exchanged in a rather unorthodox manner in order to avoid formal communication: 'Pieces of paper containing relevant information were left on office desks where the visiting Director could see them.'

The General Manager of Shell Rhodesia (G. C. Whitehead and later Robert Barrie) visited London about once a year (B12.8); there were also other contacts between British and Rhodesian oil executives, often in South Africa and, on at least one occasion, in Switzerland (see p. 72). There was obviously particularly close coordination over staffing: in June 1973, for instance, J. G. Musson (General Manager of Shell Malawi) switched posts with P. A. Reeler (a Manager of Shell and BP Rhodesia); this swap was presumably coordinated by Shell Centre in London.[20] Sources within Royal Dutch Petroleum have also claimed that Rhodesian staff attended Shell courses in Holland. As a result of these regular contacts between the oil companies in London and their Rhodesian subsidiaries, Bingham concludes, 'a good deal of information both about the Rhodesian market as a whole and the operations of the companies was gleaned,' by the head offices of Shell and BP (B12.10).

*   *   *   *

Meanwhile out in Southern Africa an important change had occurred in the supply arrangements for Rhodesia. At the end of 1971 the Total swap arrangement had suddenly been ended, and Shell Mozambique had gone back to handling Shell and BP supplies for Rhodesia. Louis Deny of Total had apparently been worried that if news of the swap had leaked out, the French company might get into trouble 'with the Black African states' in which it operated (B8.12).

The decision to end the swap was apparently taken by Louis Walker – without consultation with Shell's head office in London. Walker later justified his unilateral action by explaining that by 1971 the situation was quieter, and it didn't seem necessary to me at the time to formally notify London' (B8.13). BP gave a similar explanation: there was a 'somewhat more relaxed climate of diplomatic and political opinion' in the early 1970s, BP told Bingham, and it was presumably thought that the difference between Total South Africa and Shell Mozambique handling supplies was 'unlikely to be of real importance'.[21] Bingham himself puts part of the blame on the head offices of the oil companies and the British Government, and argues that Walker's decision was 'probably influenced by the lack of official and company concern currently apparent in relation to questions of sanctions enforcement' (B14.21).

Certainly the Conservative Government was not alone in its lack of enthusiasm for sanctions: the Western powers in general had become more 'relaxed' over Rhodesia. In March 1971 for example, the United States Government had permitted the import of Rhodesian chrome. But it was also true that the change in the supply arrangements had little practical impact on Rhodesia, in that the same amount of oil continued to reach the rebel regime. There was, however, a vital legal difference now: for when Shell Mozambique resumed handling oil for Rhodesia, it was once more clearly acting in contravention of the UK Sanctions Order.

Incredible though it seems, it took more than two years for the head offices of Shell and BP to discover that Shell Mozambique was back in the supply chain. At the time of the publication of the Bingham Report much less attention was focused on this part of the story, than the earlier discovery back in 1968 that Shell Mozambique had been busting sanctions. But the 1974 revelations are even more instructive in showing how the oil companies reacted to the Rhodesian problem.

The first news about the end of the swap was apparently not obtained until Arthur Sandford (BP) and John Francis (Shell) each separately visited South Africa early in 1974. During Sandford's visit to South Africa in February he was present at a BP meeting at which it was mentioned that Shell was proving reluctant to divulge information on what was termed a 'grey area' (B8.38). Further inquiries revealed that this euphemism referred to Rhodesia. Sandford then raised the matter with Ian McCutcheon, a British citizen who had become Managing Director of Servico (a joint Shell/BP service company in South Africa) after Louis Walker's departure in July 1973. McCutcheon confirmed that Shell and BP oil was indeed being sent to Rhodesia by Shell Mozambique.

Francis, who had apparently not yet learned about Sandford's discovery, visited South Africa the following month, and on one occasion during his trip was told that the Total swap was no longer in operation. Francis later recalled that he sat up and said, 'Ouch, what on earth did you do for that?' (B8.42). He then told senior executives of Shell South Africa: 'Look, you chaps took a good deal of trouble a few years ago to get Shell Mozambique out of the act . . . You had better take some steps to get them out of the act again, because this is going to embarrass us'.

Both Sandford and Francis did warn a number of their superiors about the renewed involvement of Shell Mozambique. But incredibly *none* of those who were given this sensitive information appears to have appreciated its significance. Sandford raised the matter with his immediate superior at BP's head office, Alan Robertson, who apparently knew little of the earlier events of 1968. Robertson then wrote to Christopher Laidlaw, the BP Managing Director responsible for Africa, to report rather casually that

Sandford had 'the impression that with the passage of time our people in South Africa may have become a little relaxed about procedures' (B8.38). Robertson also told J. de Liefde, the Shell Co-ordinator for Africa, about this news, but de Liefde claims to have no recollection of the conversation. Bingham concludes that Robertson clearly 'did not communicate any sense of urgency or alarm to the most senior levels of BP management or to Shell' (B14.4).

It is also disturbing that Sir Eric Drake, the Chairman of BP, was not informed about the situation regarding Rhodesian sales – particularly since he himself visited South Africa in February 1974. Sandford's discovery that the swap had ended had taken place just a few weeks before his Chairman's arrival, on a trip whose purpose was to prepare for the Sir Eric's visit. As Regional Co-ordinator for Africa, Sandford would presumably have been the main executive to brief the Chairman on the local situation. Yet, according to Sir Eric Drake, he did not even know about the existence of the Total swap, let alone the fact that it had been ended. 'No one broke the news to me,' the former BP Chairman admitted when the Bingham revelations were published; 'I was absolutely astonished.'[22]

Francis also took up the issue when he returned to London from his South African safari. He discussed the embarrassing news of the ending of the Total swap with his assistant (Peter Chew) and his superior (J. de Liefde). But de Liefde, says Bingham, 'did not appreciate that Shell Mozambique might be in jeopardy, nor that there was any departure from arrangements notified to HMG' (B14.4). He therefore did nothing about it, and did not inform Dirk de Bruyne, the Shell Managing Director responsible for Africa. This was unfortunate, since de Bruyne had a long experience of dealing with Rhodesia, and he at least might have appreciated the significance of the ending of the Total swap.

But it was not only news from the visits of Sandford and Francis that should have alerted Shell and BP Directors in London: revealing information on sanctions-busting was also contained in a secret BP document known as the Rounce Memorandum. This paper, unambiguously headed 'Rhodesia and Freight Services', had been prepared in February 1974 on the instructions of Denys Milne, the Managing Director of BP Southern Oil. Milne, who was a British

228

citizen, had gone out to Cape Town in May 1971 to help supervise the division of the joint Shell/BP marketing company in South Africa. Soon after his arrival, he had been warned against asking too many questions about Rhodesia: 'There were certain matters', he was told, 'into which no sensible expatriate should inquire.'[23] At this time Milne realised that Rhodesia was being supplied through Freight Services, but apparently knew little of the details and made little effort to probe further during the next two years.

Only at the end of 1973 did Milne turn his attention to Rhodesia, when John Rounce, his assistant, was commissioned to prepare a paper on this trade. The resulting memorandum, Milne later declared, was 'a typical first-class piece of work by John Rounce' (B8.40). Certainly the Rounce Memorandum, which was placed in the Sandford File, provided a concise account of the paper-chase: supplies for marketing in Rhodesia, it explained, 'are effected from South Africa primarily through "Freight Services", who act as forwarding agents, buying products from BP and Shell SA [South Africa] and reselling to the Rhodesian Government Procurement Agency Genta' (B8.39). The document went on to explain that supply volumes and prices are 'dealt with by Genta and certain representatives of the [Shell and BP] Cape Town companies under conditions of strict security, so as to avoid possible identification of the companies as direct communicants'.

The most damaging admission in the Rounce Memorandum concerned the involvement of Shell Mozambique. It explained that 'Shell Mozambique, as a UK registered company, and thus subject to Orders in Council, is not party to any sale, though the oil, theoretically in transit for South Africa, passes through their installation' (B8.39). This was not, in fact, strictly accurate, since, as Bingham discovered, supplies passed technically through the ownership of Shell Mozambique. Besides, even the handling of oil for Rhodesia would have represented a breach of the Sanctions Order.

Rounce described the Shell and BP trade with Rhodesia in explicit detail. Yet, as Bingham revealed in his typical low-key style, 'it does not appear that receipt of the Rounce Memorandum caused a great stir within BP' (B8.41). Robertson, who had failed to 'communicate any sense of urgency or alarm' after learning from Sandford that Shell Mozambique had resumed shipments to Rhodesia, appears to have shown equally little interest in the Rounce Memorandum. He explained to Bingham:

> I do remember Sandford saying to me that Rounce had prepared a very detailed note . . . which spelt out a lot of things which he hadn't understood, particularly about supplies to Rhodesia. I said, 'Is it a long note?' He said 'Yes, and it is very detailed'. I said 'Do you want me to read it?' and I think he said, 'I don't think there is any reason why you should . . . ' (B8.41).

On 12 March 1974, after reading the Rounce Memorandum, Ian McCutcheon wrote to tell Denys Milne that the arrangements were 'far too transparent and changes will have to be made' (B8.47). Eight days later McCutcheon wrote back again with specific suggestions. Shell officials would meet with Genta to 'arrange that any correspondence with Genta will not appear in our Freight Services' files or any of our official files'. He added: 'We will also arrange that there is some back-dated correspondence agreeing on volumes and prices of products moved since January 1974, so that at least from that date we can demonstrate that there have been arms length transactions with Freight Services' (B8.48).

McCutcheon went on to confess forlornly: 'It is extremely difficult to find a fool-proof arangement which will demonstrate that we have not been aware of any dealings with Rhodesia (B8.48). Nevertheless, João Vasconcelos of Shell Mozambique had been asked 'to investigate whether there are any other arrangements that could be made to mask the fact that we are selling product to Freight Services who are subsequently transporting to Rhodesia'. As the correspondence shows, the reaction of BP executives to the Rounce Memorandum was not to take steps to end supplies to Rhodesia, but to concentrate on concealing the existing trade through Mozambique more effectively.

On 15 May 1974 McCutcheon sent a note to Francis on trade with Freight Services, containing an important section on 'security'. This explained: 'No written agreements or indeed any correspondence with or related to Genta are on our official files. We have since the 1 January 1974, as can be easily demonstrated, only been dealing "at arms length" with the South African-based company Freight Services' (B8.51). Since the South African subsidiaries of Shell and BP *had* clearly corresponded with Freight Services (and probably also with Genta), the note only allows two conclusions to be drawn. First, the company might have taken a decision to destroy all incriminating documentation. Secondly, and more likely, the company might simply have set up 'unofficial' files. As is known from internal sources within the oil companies, this was a common practice: correspondence relating to Rhodesian trade was frequently sent on a 'personal' basis, often to the home addresses of the executives, in order to keep incriminating papers out of the official files.

*   *   *   *

The joint Shell/BP marketing companies in South Africa had been managed by Shell; but when these were split up in the early 1970s, BP went to considerable lengths to retain its share of the Rhodesian trade. Alan Robertson, the Regional Director of BP Trading responsible for Africa was insistent on obtaining a half share of trade in all sectors of business – but particularly so in the case of Freight Services: 'We had grave suspicions, because of the way the accounting had been done, it might . . . [have been] quietly slipped through a net and end up in Shell's hands' (B8.60). There were various categories of trade, he added, where the 'possibility (to put it crudely) of cheating' by Shell seemed to be greater. Denys Milne, who actually supervised the division of the joint companies for BP, explained: 'Irrespective of politics, and one might say law, the fact is, here was a piece of business of South African origin with a label on it, and my concern was to get my half in my name so that nobody could touch it'.

The renewed involvement of Shell Mozambique in handling Rhodesian supplies caused some personal concern to a number of Directors who were presumably worried that they might be contravening the Sanctions Order. Soon after John Francis had learned that Shell Mozambique was again busting sanctions, there took place a revealing exchange of correspondence, which was reminiscent of Dirk de Bruyne's attempt back in 1968 to get 'assurances' from Louis Walker

231

over sanctions (see p. 189). On 3 June 1974 Francis wrote to Ian McCutcheon asking for an assurance from Shell Mozambique that 'products emanating from the Company's installations are not reaching Rhodesia' (B8.56). It was a rather strange request, considering that Francis had just learned that Shell Mozambique *was* handling Rhodesian oil, and Francis himself admitted that the letter was 'partly for the record'. Bingham notes that the letter certainly struck 'a suprising note'. It was not until over three months later that the answer came back. 'Upon reflection,' McCutcheon replied, 'you will appreciate it is quite impossible for me to give you the unqualified assurance you have requested' (B8.57).

By the spring of 1974 the renewed involvement of Shell Mozambique in handling Rhodesian oil was known to the three British Directors of Shell Mozambique (Francis, McCutcheon, and Sandford), and probably to at least six other British executives of Shell and BP (Chew, Laidlaw, Milne, Robertson, Rounce, and Trechman). It is also difficult to believe that Wayne Templer (who became Chairman of BP South Africa and a Director of Shell Mozambique in September 1975), and Michael Savage (who replaced Sandford as BP Regional Co-ordinator for Africa and became a Director of Shell Mozambique in October 1975) would not have known that the joint Shell/BP company in Mozambique was supplying Rhodesia.

Thus, as we have seen, a number of the BP Directors who discovered that Freight Services was a euphemism for Rhodesia tried to ensure that Shell did not 'cheat' by letting more than its share of the Rhodesian trade 'slip through a net'; other oil executives were concerned to seek for ways 'to mask the fact that we are selling products to Freight Services, who are subsequently transporting to Rhodesia'; one Shell executive, after discovering that Shell Mozambique *was* handling Rhodesian supplies, wrote to ask for 'assurances' that this was *not* the case. But the one thing that the oil men seem to have failed to do was to take any meaningful steps to ensure that Shell Mozambique was not contravening the UK Sanctions Order. There certainly seems to have been absolutely no attempt made to actually cut off Shell and BP sales to Rhodesia.

*   *   *   *

Sandford and Francis, the two London-based executives who first discovered that Shell Mozambique was again involved in handling Rhodesian trade, appear to have been particularly lax over seeing that the UK Sanctions Order was being respected. Bingham argues that they should have ensured 'that Shell Mozambique had been removed from the chain of supply to Freight Services (or, if it had not, to seek some alternative expedient)' (B14.23). (Bingham, however, does not state what sort of 'expedient' would have been acceptable.)

Sandford, who was on a second trip to Southern Africa in October 1974, visited Lourenço Marques, and thus had a good opportunity to investigate the role of Shell Mozambique at first hand. But apparently he was 'struck by the absence of rail tank-cars anywhere to be seen, and was given to understand that they were now all on the Beit Bridge line', the direct rail link that had just been opened from South Africa to Rhodesia (B8.55). Sandford could not have had his eyes completely open. For at this time only a small part of Rhodesia's supplies were sent on the new Beit Bridge rail link, and the main route continued to be the line from Lourenço Marques. Bernard Rivers, who had visited the marshalling yard just a few days before Sandford, had seen rows of Rhodesia Railways tank-cars bound for Rhodesia. Sandford, as Bingham remarks, must have been 'easily satisfied' (B14.23). Bingham concludes that 'there is no evidence that Sandford, the BP Regional Co-ordinator for Africa, 'pursued the matter in any way, or that anyone in BP pursued the matter with him' before his retirement in September 1975 (B8.72).

Francis was equally ineffective. In March 1974 he had told the local management of Shell in South Africa that immediate steps should be taken to remove Shell Mozambique from the chain of supply to Freight Services. Correspondence he received in May and September 1974 showed that this had *not* been done. But presumably comforting himself with the news of the completion of the Beit Bridge rail link and with Sandford's comments that this route was now supplying Rhodesia, Francis 'ceased to pay much further attention to the position of Shell Mozambique' (B8.73).

Francis later received a crucial letter dated 11 February

233

1975 from Kenneth Geeling, Walker's replacement as head of Shell South Africa, which explained that Shell Mozambique was still handling some oil for Rhodesia. Bingham therefore concludes that after February 1975 Francis 'must have known that Freight Services trade was continuing to involve Shell Mozambique' (B8.74). Yet in spite of this he does not appear to have taken any action to pull Shell Mozambique out of the supply chain. Francis, Bingham comments gently, should not 'have let the matter rest' (B14.23).

By the mid-1970s the head offices of Shell and BP were in a better position than ever to take action to end Shell Mozambique's involvement in Rhodesian trade. In September 1973 Gavin Trechman, a British BP executive, was sent out to Lourenço Marques as a Senior Manager of Shell Mozambique and two years later was promoted to General Manager of the company, reporting direct to London. This was exactly what BP had called for back in 1968, when it had been discovered that Shell Mozambique had been busting sanctions; but on that occasion the proposal had been vetoed by Shell (see p. 188).

The decision to appoint Trechman as General Manager, however, was made purely as a result of BP taking over from Shell as the manager of Shell Mozambique in September 1975; the move had nothing to do with Rhodesian considerations. Indeed, incredible though it seems, Trechman apparently was not even given instructions regarding Rhodesia when he took up the post. He confessed to Bingham that when he was in Mozambique he 'never gave any thought to the fact that we might be contravening the laws of another country' (B8.64). This was an astonishing statement, since it suggests that Trechman may not even have been aware of the existence of the UK Sanctions Order.

Certainly it must have been clear to Trechman as General Manager that nearly half the oil which his company handled was destined for Rhodesia. Shell and BP documents often referred to 'Freight Services' as if it was a country – and the identity of that country was surely unmistakable. Every week Shell South Africa would send Shell Mozambique a telex giving the loading programme for the next seven days. These requirements were always split into four 'countries' – South Africa, Botswana, Swaziland . . . and Freight Services. The stock records of Shell Mozambique only contained two categories of exports: the Southern African Customs Area (South Africa, Botswana, and Swaziland) and what was euphemistically called 'FS Exercise'. Even a brief glance at a map would have showed that if products sold in bond to Freight Services at Lourenço Marques were not bound for the Southern African Customs Area, then Rhodesia was the only possible destination. But despite the evidence that supplies to Freight Services were destined for Rhodesia, it does not seem that Trechman made any attempt to investigate the matter – let alone end these sales.

*   *   *   *

A few months after the Labour Government came back to power in 1974, Foreign Secretary James Callaghan told Parliament: 'We have stepped up our sanctions surveillance effort.'[24] But in spite of this the Government apparently never discovered from its own intelligence services that the Total swap had ended, and that Shell Mozambique was again handling Rhodesian supplies.

For the British Consuls at Lourenço Marques – Alan Free-Gore (1971–73) and Stanley Duncan (1973–75) – this should have been an easy task, the simplest step being merely to ask the British head of Shell Mozambique, Gavin Trechman, whether the swap was still in operation. In fact Trechman had been approached by a consular official in February 1974, who had inquired about oil shipments to Rhodesia; but the Shell Mozambique executive replied that he had only just arrived in Lourenço Marques, and was unable to provide useful information (B8.81). No reason has been given why Trechman should not have been asked again a few months later. It is also unclear why intelligence agents in Southern Africa were not able to discover that the Total swap had ended – and what they *were* doing if they were not reporting on sanctions-busting.

The return of the Labour Party to power in March 1974 might have been expected to lead to a change in Britain's attitude to sanctions. Lord Elwyn-Jones, the new Lord Chancellor, later claimed that 'when in 1974 a Labour Government returned to office, particular efforts were made – with some success – to encourage a more determined application

235

of sanctions by our international partners'.[25] But this merely repeated a now-familiar theme – that it was other countries which were busting sanctions, not Britain itself.

Dr David Owen has told Parliament that he had 'found no evidence to indicate that Ministers [of the incoming Labour Government] were told about the swap arrangements.'[26] Neither of the two politicians who clearly had known about the swap were in the new Cabinet: Thomson was based in Brussels as the EEC Commissioner, and Stewart remained a backbencher in the House of Commons. Wilson, who was now Prime Minister again, has of course claimed that he had known nothing about the swap arrangement. It is not known whether or not James Callaghan, who took over as Foreign Secretary, had been told. But David Ennals, the Foreign Office Minister of State responsible for Africa, has claimed that he was 'never informed that Shell Mozambique had been busting sanctions in the 1960s, and that a swap arrangement with Total had subsequently been set up to supply Rhodesia'.[27] Ennals added: 'Had I heard any such reports, I would have called for an immediate inquiry.' Apparently, no questions were asked of the civil servants responsible for dealing with the oil embargo at the Foreign Office or the Ministry of Technology, and no information was volunteered by them. The skeletons were kept safely in the cupboard.

The British Government again seems to have simply assumed that British companies always played by the rules, and that if international sanctions were being defied, then wicked foreigners were the culprits. A bizarre incident a few months after the Labour Party came to power illustrated this recurring theme. On 7 May 1974 Britain sent a formal note to the UN Sanctions Committee, stating that it had discovered that a Rhodesian firm called Master Stores was 'actively seeking to purchase crude oil'.[28] Britain went on to request UN members to 'advise any exporters of crude oil operating in their territories of the activities of Master Stores'. But if diligent British investigators had managed to uncover the fact that a Rhodesian company as obscure as Master Stores was attempting to enter the trade, why was it that they were unable to discover that the British firm of Shell Mozambique had actually supplied half of Rhodesia's oil for most of the period since UDI?

During 1974 there was an increasing volume of evidence that Shell and BP were involved in sanctions-busting. Part of this was gathered by a man with the appropriate-sounding name of Ernest Fellowes, who had been living in Southern Africa and had observed that the railway line from Lourenço Marques to Rhodesia was jammed with long lines of tank-cars. As a loyal citizen he twice reported his discovery to the British Consulate at Lourenço Marques. But the dusty response he got was simply, 'Thanks very much, but we know all about this.'[29] Fellowes' reaction was understandable: 'I expect these boys have got agents all over Rhodesia. Better leave it to the professionals.' Nevertheless, on his return to London, he presented a 12-page report on sanctions-busting to the Rhodesia Desk of the Foreign Office.

By this time the Portuguese coup and the growing moves toward Mozambican independence had also made it easier to investigate sanctions-busting. Surely this important opportunity should have been used to examine once more whether sanctions could be made more effective?

David Martin, the *Observer*'s African correspondent, reported on 29 September 1974 that British oil companies (among others) were 'operating through intermediaries' to supply Rhodesia from Lourenço Marques. The *Observer* did not provide further details, and Freight Services was not named as the intermediary; but the article should nevertheless have encouraged the Government to investigate the possibility that Shell Mozambique might be busting sanctions. Unfortunately by now Britain seemed to have lost all interest in the subject.

It is remarkable that after Thomson's meeting with Shell and BP in February 1969, there were no further contacts between the Government and the oil companies over Rhodesia for more than seven years. During this period there were various changes of Government in Britain – the Conservatives came in in 1970, and Labour was back again in 1974; there were also further attempts to find a solution to the Rhodesian problem, including the 1971–2 proposals; Rhodesia continued to be one of the main issues facing the Foreign Office. But in spite of the Government's controlling interest in one of the oil companies concerned, BP and Shell's activities in Southern Africa seem to have been of no interest to Westminster or Whitehall. Sir Robin Brook, who served as a Government-appointed Director of BP between 1970 and 1973 has claimed that he was never told about the existence of the Total swap.[30] His successor, Lord Greenhill, has said that until 1976 'the question of sanctions was not mentioned by the Government to me or to my other Government Director colleagues . . . although I was in close contact with Whitehall'.[31]

Ministers repeatedly told Parliament that further steps were being taken to tighten sanctions; and Britain periodically lectured the United Nations on the need for other nations to follow the UK's tough stand. But meanwhile the British Government made absolutely no effort to determine whether the country's two largest companies were respecting sanctions. The Rhodesian charade continued.

*　　*　　*　　*

There was only one moment when it looked as if the oil embargo might actually starve Rhodesia of fuel. This was in October 1973, as a result of the global oil crisis sparked off by the Arab-Israeli war. Suddenly production was cut, a number of consuming countries were embargoed for political reasons, and prices leapt upwards. The oil crisis demonstrated that Third World countries could wield a devastating economic weapon for political ends: many African countries backed the Arab struggle against Israel, and in return the Africans won greater Arab support against minority rule in Southern Africa. On 28 November 1973 a summit conference of Arab nations imposed a complete oil embargo against Rhodesia, South Africa and Portugal.

Initially the Arab embargo caused considerable concern in Rhodesia. Petrol rationing was reintroduced on 1 February 1974; a 'Don't Drive Rhodesia Dry' campaign was launched, and posters suddenly appeared in the streets of Salisbury exhorting motorists to 'Take home a secretary' or 'Stay at her place tonight'. There was not only the problem of finding oil – but also of paying for it, for the leap in the world price of oil, from just over £1 to £5 a barrel within four months, gave a drastic jolt to the Rhodesian economy. Rationing and the jump in price had the effect of reducing consumption considerably over the next year. But after a few difficult weeks in late 1973 and early 1974, the Rhodesians discovered that they were able to import all the oil they could afford.

Sonarep was the only company which had major problems in supplying Rhodesia. The refinery had great difficulty in obtaining crude oil, and its exports to Rhodesia were therefore temporarily halted at the end of 1973. This caused some concern to the South African Government, which was worried that the oil companies might actually send too much oil to Rhodesia and reduce supplies for the South African market. The companies were therefore summoned to a meeting with the Ministry of Commerce on 31 October 1973 at which they were told that they 'should not make up for additional shortfalls of supply of Freight Services resulting from the inability of Sonarep to supply' (B8.77). This suggests that the oil companies were far from being the unwilling tools of the South African Government in busting sanctions, as they had claimed in their references to the local 'conditional selling' laws. The South African Government feared that they were actively trying to *expand* their markets in Rhodesia in order to take over Sonarep's share.

The South African subsidiaries of the oil companies played a key role in ensuring that the Arab embargo against Rhodesia proved ineffective. In turn it was the international oil companies themselves which kept South Africa afloat by shipping sufficient oil for its own use and for export to Rhodesia. When Sir Eric Drake, Chairman of BP, visited South Africa in March 1974 he claimed that the oil companies, and BP in particular, had 'intentionally set out to thwart Arab attempts at enforcing embargoes on countries like South Africa'.[32]

Until 1973 around half of South Africa's oil had been imported from Iran. The remainder, which mainly came from Iraq, Saudi Arabia, and Qatar, was cut off after the Arab embargo. But the oil companies managed to 'juggle' supplies: Iranian oil was sent to South Africa, and other countries which had previously received Iranian oil were supplied with Arab oil. As the Johannesburg *Financial Mail* explained: 'There can be no greater blessing for South Africa – apart from the fact that Iran is well disposed – than that the oil business is still largely in the hands of international companies with no discernible leanings of excessive patriotism.'[33] It was a remark that was truer than its author could ever have known.

NOTES

1 BBC Television, 19 September 1978.
2 Hansard, Commons, 8 November 1978, col. 994.
3 *Daily Mail*, 29 August 1978.
4 *Guardian*, 12 September 1978.
5 *Guardian*, 28 August 1978.
6 *Guardian*, 16 September 1978.
7 *Guardian*, 16 September 1978.
8 *Guardian*, 16 September 1978.
9 BBC Radio, 19 September 1978.
10 *Guardian*, 15 September 1978.
11 *Observer*, 10 September 1978.
12 Letter from Jock Davidson (Freight Services, Lourenço Marques), 5 November 1970 addressed to the Mozambique offices of Caltex, Mobil, Shell, Sonarep, and Total.
13 Granada Television, *World in Action*, 7 August 1978.
14 Granada Television, *World in Action*, 7 August 1978.
15 *To The Point International*, 27 February 1978.
16 Interview, 13 July 1978.
17 Rapport, 12 December 1971.
18 *The Oil Conspiracy*, p. 25.
19 Memorandum from Cecil Low (BP South Africa) to Denys Milne, 6 May 1974 (unpublished).
20 Letter from Douglas Gash (Rhodesia Oil Company) to Joao Bartolo (Sonarep, Lourenço Marques), 26 March 1973 (unpublished).
21 BP Submission to the Bingham Inquiry, 27 September 1977, p. 82.
22 *Guardian*, 28 August 1978.
23 BP Submission to the Bingham Inquiry, 27 September 1977, p. 83.
24 Hansard, Commons, 8 November 1974, col. 1400.
25 Hansard, Lords, 9 November 1978, col. 432.
26 Hansard, Commons, 7 November 1978, col. 705.
27 Interview, 27 March 1979.
28 UN Sanctions Committee 7th annual report, pp. 118–9.
29 *Guardian*, 14 September 1978.
30 *Guardian*, 9 November 1978.
31 Hansard, Lords, 9 November 1978, col. 522.
32 *Rand Daily Mail*, 5 March 1974.
33 *Financial Mail*, 7 December 1973.

12   *The South African Lifeline*

March 1976–1979

The coup in Portugal on 25 April 1974 was disastrous news for the Smith regime. It rapidly became clear that Mozambique would win its independence, and that a Government led by the FRELIMO guerrilla fighters would be likely to cut Rhodesia's vital rail outlets to the sea. The oil companies decided to make immediate contingency plans. On 4 July 1974, just two months after the coup, Kenneth Geeling of Shell and William Beck of Mobil called on the South African Secretary for Commerce on behalf of all the oil companies operating in the Republic. The main items on the agenda were 'Extraneous Demand' – a euphemism for Rhodesian trade – and the 'uncertainties' created by events in Portugal. The Shell and Mobil executives stressed that political developments might make it impossible to rail oil supplies from Lourenço Marques to 'our northern neighbour'. In this case, Geeling and Beck inquired, what would be the Government's attitude 'if all these supplies were to be switched to South African sources' (B/A 281)?

A new direct rail link between South Africa and Rhodesia was already nearing completion. The existing railway from Johannesburg ran to the Beit Bridge border, but on the Rhodesian side there was a 100-mile gap between there and Rutenga, a small town on the line from Lourenço Marques. Work on completing the missing section had started in 1972, but construction was speeded up after the Portuguese coup.

Shell and Mobil therefore suggested that plans should be made to divert Rhodesia's oil supplies to the new Beit Bridge line. Secretary for Commerce Joep Steyn replied that this matter was too important for him to settle alone, and the oil companies' request would have to go to Cabinet. His fear, however, was that if Rhodesian supplies went via

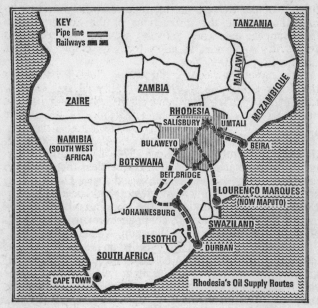

Fig. 6

At UDI Rhodesia imported crude oil – this was transported by pipeline from Beira and refined at Umtali. Since UDI all oil supplies have been in refined form. From 1966 until the closure of the Mozambique–Rhodesia frontier in March 1976, the oil was mainly sent on the railway that runs northwards from the port of Lourenço Marques. Since then all Rhodesia's oil requirements have been imported from South Africa, mainly on the railway route from Johannesburg via the Beit Bridge border.

Beit Bridge 'then it would become even more apparent to the world at large that South Africa remained the only culprit; this could turn the full spotlight of international wrath against South Africa, with serious consequences' (B/A 281). Steyn added that even Iran, which provided most of South Africa's oil, 'might be embarrassed in such a situation.' This was a revealing discussion: it suggests that

242

the oil companies were not mere pawns, who were supplying Rhodesia on the instructions of the South African Government, but were actively pressing the somewhat reluctant Government to allow them to supply Rhodesia directly from South Africa.

This impression seems confirmed by Geeling's comment that he and Beck thought that the South African Government would in the end 'be almost compelled to assist, because of the strategic and emotional significance of the matter' (B/A 281). In other words, the oil companies believed that the South African Government would *have* to permit direct oil shipments to Rhodesia, due to both the strategic importance of keeping the Smith regime in power and the sympathy most white South Africans had for the settler community in Rhodesia. It is clear, at least from this particular meeting, that it was the oil companies which were making the demands and that the South African Government was initially reluctant to give in to them.

The South African subsidiaries of Shell and Mobil then came up with a proposal for a new swap arrangement. 'If all supplies were to emanate from South Africa,' they explained, then 'consideration should be given to Sasol [the state oil corporation] directly handling all these supplies and balancing through inter-company exchanges' (B/A 281). This proposal must have been approved subsequently by the South African Cabinet, for a few months later Shell South Africa took the initiative and began negotiations with Sasol. Shell's suggestion was that Sasol should supply Rhodesia, and that in return the international oil companies should increase their sales in the domestic South African market. This was to be the basis of the Sasol swap – an arrangement, which when publicly revealed four years later, was to cause a major row in Britain.

It took many months to negotiate details of the swap, which was a considerably more complicated arrangement than the earlier swap with Total. Sasol's main source of oil was the new National Refinery (Natref), which had been opened three years earlier at Sasolburg, thirty miles south of Johannesburg. Sasol itself held a controlling 52.5 per cent shareholding in Natref, with Total (30 per cent) and the National Iranian Oil Corporation (17.5 per cent) holding the remaining shares. When the refinery had come on stream

243

in 1971, a complicated agreement had been hammered out under which Sasol was given the right to supply part of the South African market in what became known as the Natref Area. But by the end of 1974, after the completion of the new direct rail link at Beit Bridge, the Natref refinery became the most convenient source of oil for Rhodesia. The Shell/BP and Mobil refineries at Durban were 700 miles from Beit Bridge, over twice as far as the Natref refinery. There would therefore be considerable savings in transport costs if Rhodesian supplies were sent direct from Natref.

At the end of 1974, just a few weeks after the completion of the new rail link, Sasol despatched its first exports to Rhodesia. At this time, however, the refinery had insufficient output to sell to Rhodesia without cutting back on its domestic sales in the Natref Area. This meant that as the oil companies lost part of their share in the Rhodesian market, so they increased their sales inside South Africa.

By late 1975 Natref's capacity had been expanded, so that the state-controlled refinery was now able to sell to Rhodesia *and* maintain its market in the Natref Area. But at this point the international oil companies became worried that Sasol would encroach on their trade. As Bingham reported, they were anxious not to lose the right to refine 'the volume of crude oil roughly equivalent to the Freight Services business' (B13.5). Shell and BP therefore hastily arranged another meeting with Sasol to try to prevent the South African oil corporation from supplying Rhodesia at their expense.

Another complicated arrangement was worked out over the next few months. Sasol had already begun providing about 15 per cent of Rhodesia's requirements, but the question now arose as to what should happen to the 85 per cent of the trade still in the hands of the international companies. Eventually it was agreed that Sasol should ultimately provide all of Rhodesia's requirements of the main oil products (such as petrol and diesel), and in return the international oil companies would be given the right to supply the Natref Area of South Africa with additional quantities of oil products equivalent to 85 per cent of Freight Services' purchases.

A note prepared by E. J. Bonds from Shell Centre in London explained that the complicated Sasol swap was considered quite satisfactory. 'It was originally feared that Shell's substantial share of this [Rhodesian] business would be lost,' Bonds explained, 'but arrangements have now been made which will mean that Shell will still be able to supply its share' (B13.9). The international oil companies thus expanded their share of the internal South African market at the expense of Sasol, while Sasol began handling the politically sensitive supplies for Rhodesia.

On 3 March 1976, after a Rhodesian border raid into Mozambique, the newly independent Frelimo Government in Mozambique finally closed its frontier with Rhodesia. At this time most of Rhodesia's oil supplies were still being despatched from Lourenço Marques (although Sasol had already begun sending small quantities on the new Beit Bridge line). The border closure created one immediate problem: a number of Rhodesia Railways tank-cars, which had been loaded by Shell Mozambique, were stuck on the Mozambican side of the frontier. On 3 March these were seized by the Mozambique Government and were brought back to Lourenço Marques marshalling yard as incriminating evidence that the company had been busting sanctions. Many months later Shell Mozambique negotiated with the Mozambique Government to buy back the oil.

*   *   *   *

Political developments in Mozambique had already had an important impact on the Rhodesian situation. Mozambican independence and the subsequent closing of the Rhodesian border made the Smith regime even more isolated; and freedom for Mozambique also gave the nationalist movements an important new base for their guerrilla fighters. Back in the late 1960s both the Zimbabwe African People's Union (ZAPU) and the Zimbabwe African National Union (ZANU) had turned to the gun. After UDI it had quickly become apparent that sanctions would probably never topple the Smith regime and that Britain and the other Western powers simply lacked the determination to make sanctions bite. Reluctantly, the Rhodesian nationalists realised that freedom would only be achieved by armed conflict.

After the Portuguese coup, with independence for

Mozambique clearly on the way, Smith realised that he had little choice but to try to seek some accommodation with the African nationalists. In December 1974 an attempt was therefore made to reach a ceasefire in the escalating guerrilla war. Although the ceasefire never became effective, the détente exercise did have an important result: the release of many of the leading nationalists from detention – including figures such as Joshua Nkomo, the Reverend Ndabaningi Sithole, and Robert Mugabe.

Formal negotiations between Smith and the nationalists leaders were held at Victoria Falls in August 1975. These soon broke down, however, as did the fragile unity between the nationalists. Bishop Muzorewa emerged as the head of the United African National Council; Sithole and Mugabe remained outside the country with ZANU; and Nkomo returned to Salisbury as head of ZAPU.

Four months later Smith made a further attempt to reach a settlement. Talks continued between Smith and Nkomo until March 1976, when they finally broke down. Meanwhile the guerrilla war, led by ZANU forces based in Mozambique, was causing a rapid deterioration in the security situation inside Rhodesia; it was on 3 March 1976, after a retaliation raid into Mozambique by Rhodesian forces, that the border between the two countries was closed.

South Africa, which as a result of the border closure had now become Rhodesia's only outlet to the sea, was the rebel regime's lifeline with the outside world – and felt increasingly exposed. As the Secretary of Commerce had predicted, his country was now in 'the full spotlight of international wrath'.

Within a few weeks of the border closure considerable congestion developed on the South African railway system, severely affecting Rhodesia's trade. Oil traffic was held up, and at one point, in June 1976, Rhodesia's oil stockpile was apparently down to just under twenty days' supply.[1] It was widely believed that this 'congestion' had been partly created by the South African authorities to remind the Rhodesians of their dependence on their southern neighbour.

The United States Government was becoming increasingly concerned at the unstable situation in Southern Africa. Secretary of State Dr Henry Kissinger therefore made an important initiative to work for a Rhodesian settlement. In September 1976 he met Smith in Pretoria and warned him that if majority rule was not conceded, then Rhodesia would come under increasing international pressure. A few days later Smith told an astonished Rhodesian public that he had agreed to majority rule within two years – an abrupt turnabout for a leader who only a few months earlier had predicted that Rhodesia would not see African rule 'in a thousand years'. Smith later explained that under the threatened international pressure, Rhodesia 'would not probably continue to survive, and therefore we made this dramatic change'.[2]

Smith's belated concession was to lead to yet another attempt at a settlement, this time at a round-table conference at Geneva. The Geneva Conference was convened on 28 October 1976 to negotiate the formation of an interim administration to govern the country before full independence under majority rule was achieved. But within several weeks, the Geneva Conference foundered on the wide divisions between the Smith regime and the nationalists.

During 1977 international efforts to reach a settlement continued. The new Foreign Secretary, Dr David Owen, worked closely with US Ambassador to the UN Andrew Young on new proposals. The Anglo-American Plan, which was formally announced on 1 September 1977, involved the temporary hand-over of power to a British Resident Commissioner, a six-month transitional period leading to free elections under universal adult suffrage, and the formation of a national army based on the guerrilla forces. Right from the start, however, both the Smith regime and the African nationalists reacted cautiously to the Anglo-American Proposals.

* * * *

It was during this period of intense diplomatic activity over Rhodesia that the oil companies first came under strong attack over allegations of sanctions-busting. *The Oil Conspiracy* was published in June 1976, just three months after the closure of the Mozambique-Rhodesia border; to the relief of Shell and BP the report contained only incidental references to the British-owned companies. Nevertheless Frank McFadzean, the Chairman of Shell, wrote to Michael

Palliser, who was now permanent head of the Foreign Office, to claim that no Shell companies were busting sanctions. The British Government then told the United Nations that it 'had accepted the assurances given by Shell and BP that neither they nor any company in which they have an interest have engaged, either directly or with others, in supplying crude oil or oil products to Rhodesia'.[3]

Yet despite the storm of denials which followed, and which culminated in Britain's misleading statement to the UN (see p. 43), the attacks on the oil companies rapidly intensified during the early months of 1977. The collapse of the Geneva Conference had led to an escalation of the guerrilla war inside Rhodesia and to nationalist calls for further pressure on the Smith regime. President Jimmy Carter, who took over as the US President on 20 January 1977, was anxious to see a settlement in Rhodesia. The publication of my report on *Shell and BP in South Africa* on 1 March, the subsequent announcement of legal action by Lonrho and the Zambian Government, and finally the establishment of the Bingham Inquiry on 8 April all put the oil companies in the limelight.

In April 1977 the Chairman of Shell, Michael Pocock, went on a secret visit to South Africa to investigate the company's involvement in Rhodesia's trade. Pocock was greeted on his arrival by a letter from his company's South African lawyers, warning him that he could face arrest as a 'foreign agent' if he attempted to elicit information on oil supplies (B13.28). Although in the end he therefore seems to have made little attempt to investigate this issue, the fact of his visit illustrated the sensitive nature of the allegations against Shell.

BP proved rather more effective in its belated attempt to sever its Rhodesian links. In June 1977 Geoffrey Butcher, the BP Trading Director responsible for Africa, visited South Africa to investigate whether oil was still being supplied to Freight Services, and if so whether these sales could be halted. He met the Minister of Economic Affairs in Pretoria, a sign of the importance which the Government attached to the matter, and issued a blunt threat. Butcher told the Minister that unless all trade with Freight Services was stopped, and confirmation received by BP's head office in London, then oil supplies to BP South Africa might well have to be withheld. The meeting, BP later admitted to Bingham, was 'somewhat tense'.[4]

Initially, BP South Africa refused to provide the guarantees which head office had requested, and on 12 July 1977 an instruction then had to be issued by BP in London banning the sale of certain specialised oil products which appeared to be still supplied to Freight Services by BP's own South African subsidiary. The loading of some consignments was halted, and arrangements were made to divert shipments already on the high seas. BP South Africa immediately raised the matter at a high level with the South African Government, where it was subsequently taken up by Prime Minister John Vorster himself. Nine days later BP South Africa climbed down: a guarantee was given to BP in London that no further sales would be made to Freight Services or to any other customers who were believed to be acting for Rhodesia.

It is strange, however, that BP's head office in London continued to list Freight Services' Cape Town telephone number in its internal directory. The Freight Services' number was still included in the September 1978 edition of the directory, published over a year after BP claims to have ended all sales to the forwarding company.

BP had now done what it had claimed for over a decade was impossible. Hitherto the company had argued that South African legislation on 'conditional selling' made it illegal to cut off sales to Freight Services. Yet the fact that this step was taken in July 1977, with no legal action following, confirms that the oil companies had been 'sheltering' under a rather dubious interpretation of South African law.

One of the most surprising aspects of BP's decision to force its South African subsidiary to cut off sales to Freight Services is that this action was not taken under pressure from the British Government. In fact, only four months earlier the Foreign Office had apparently advised *against* the plan. After the publication of my report on *Shell and BP in South Africa*, the Deputy Chairman of BP, Montague Pennell, called on Sir Anthony Duff at the Foreign Office to talk with the senior official dealing with African affairs. Pennell apparently canvassed the idea of instructing BP South Africa to cut off supplies to Freight Services, but BP later told Bingham that Sir Antony Duff vetoed the plan, saying that the Government was 'formulating its policy and such a move would be premature.'[5] It seems incredible that more than eleven years after UDI the British Government was still 'formulating' its position over the oil embargo; but what is even more surprising, if BP's account is correct, is the fact that the Foreign Office positively *discouraged* an oil company initiative to make sanctions more effective.

News of BP's tough action over sales to Freight Services was regarded as highly confidential; it would have been thought extremely embarrassing if BP's South African subsidiary had been seen to be responding so blatantly to foreign pressure. Nevertheless, Shell realised that it could find itself in a difficult position if it did not take its own initiative, and in August 1977 a team of Shell lawyers therefore flew out to South Africa to seek a similar guarantee. It was not until 30 September, however, that an assurance was received: no products provided by Shell International Petroleum would now be supplied to Rhodesia or to any person believed to be acting on behalf of a Rhodesian purchaser – unless to refuse to do so would involve a breach of the 'conditional selling' law (B13.31). This last clause, it could be argued, invalidated the undertaking; and it remains unclear why Shell was not able to obtain the absolute assurance which BP had received only two months earlier.

Why was it, then, that the British oil companies suddenly took the belated decision in 1977 to try to cut off supplies to Freight Services? By this time Rhodesia's requirements of the main oil products were being supplied by Sasol under the swap arrangement, and this meant that sales to Freight Services by the South African subsidiaries of Shell and BP consisted only of certain specialised oil products. The financial loss from ending this trade was therefore relatively small compared to what it would have been before the closure of the Mozambique-Rhodesia border the previous year, when Shell and BP were supplying the whole range of oil products.

By 1977 the oil companies were coming under considerable international pressure over sanctions-busting, and there had been widespread allegations that Freight Services was being used as a channel to supply Rhodesia. The British oil companies must therefore have concluded that the profits accruing from the sale of specialised oil products to Freight

250

Services were not commensurate with the complications which this was creating outside Southern Africa.

*   *   *   *

Even after Shell and BP had cut off sales to Freight Services, however, the South African subsidiaries of the two companies continued to participate in a formal arrangement to supply their Rhodesian companies. Under the terms of the Sasol swap, both Shell and BP were regularly notified of the quantities of oil supplied to Freight Services on their behalf. Shell and BP would then provide matching quantities of oil products to the Natref Area in South Africa.

It was not until 1978 that the existence of the Sasol swap was publicly revealed. On 10 September 1978, just nine days before the publication of the Bingham Report, the *Sunday Times* carried a front-page story with the banner headline, 'Shell and BP are still helping Rhodesia'. The article, written by Bernard Rivers, *Sunday Times* reporter Peter Kellner, and myself, exposed the existence of the swap arrangement with Sasol.

Shell refused to comment, but BP indignantly denied the allegations: 'There is no swap arrangement,' John Collins, the head of BP's Public Relations told the *Sunday Times*. on 11 September *The Times* reported an ingenuous piece of disinformation from a BP spokesman: there was indeed 'a very complicated marketing agreement between BP and Sasol . . . It is complicated because of the trouble that has been taken to prevent any of the BP oil reaching Rhodesia . . . These arrangements . . . are very difficult to understand for anyone not intimately involved.'[6]

The avowed ignorance of the oil companies of Sasol's role in handling Rhodesian trade is difficult to understand, in view of the fact that most of the senior executives dealing with Southern Africa probably knew of the existence of the Sasol swap. A copy of the record of the Shell/Mobil meeting with the South African Secretary for Commerce had been sent to Denys Milne, the British Managing Director of BP Southern Oil; Milne had then forwarded the record to Arthur Sandford in London; and E. J. Bonds, of Shell Centre, knew the precise details of how the swap arrangement operated. It is also likely that the existence of the Sasol swap was known to Wayne Templer (Chairman of BP South Africa), Ian McCutcheon (Managing Director of Servico, the joint Shell/BP service company in South Africa), and John Francis (Shell Area Co-ordinator for Africa).

But there was one fact which gave the head offices of Shell and BP absolutely no excuse for not knowing about the Sasol swap. In July 1978, two months earlier, the head offices of the British oil companies had been given, in confidence, a draft copy of the Bingham Report. This had contained a chapter on 'The Pattern of Supply to Rhodesia since March 1976: Shell and BP Involvement'. This particular section, dealing as it did with the current situation, should surely have been studied particularly carefully when is was received, and after the *Sunday Times* report one would have expected that the oil companies would again have turned to these important pages before making any public comment on the allegations. Yet the fact remains that initially the oil companies gave the impression that there was no truth at all in the newspaper report.

On the Sunday on which our article had appeared Dr Owen telephoned the Chairman of BP in an angry mood. Only five days earlier the Foreign Secretary had told the Royal Commonwealth Society that since the summer of 1977 'no oil – I am assured – from British companies . . . is reaching Rhodesia'.[7] The Sasol swap, Dr Owen later admitted, was 'totally incompatible with the spirit, if not the letter, of the assurances' he had received.[8] The oil embargo had become a sensitive international issue, and the Government had at least to be *seen* to be taking a tough stand. Dr Owen therefore summoned Shell and BP executives to the Foreign Office on two occasions during the week following publication of the *Sunday Times* report.

The Government's shareholding in BP made the position particularly delicate. BP Chairman Sir David Steel, fearing that a further scandal might lead to pressure for the full nationalisation of BP, ordered an immediate investigation. Geoffrey Butcher, the BP Trading Director responsible for Africa, was sent out once more to South Africa. He learned that the Sasol swap had indeed existed. But he flew back to London with a written assurance from Wayne Templer, the British Chairman of BP South Africa, that the Sasol swap had been terminated in July 1977, at the same time that sales to Freight Services had been ended.

When the Bingham Report was finally published on 19 September 1978 it confirmed that a swap arrangement with Sasol had indeed been established to supply Rhodesia. But Bingham suggested that the swap was still in operation: 'So far as we have been able to ascertain,' he concluded, 'this situation has continued up to the present' (B14.4). Immediately after the Bingham Report was published, however, BP issued a statement referring to Bingham's description of the Sasol swap, and claiming that 'not even those arrangements now exist'.[9] It went on to describe the *Sunday Times* article as both 'offensive and defamatory'.

Shell also issued its own denial of the Sasol swap. It claimed that the company had 'received fresh assurances which enable us to state categorically that there is no arrangement of any kind between Shell South Africa and any other company for Shell South Africa to share in or be "compensated" for Rhodesian business'.[10]

Meanwhile back in Cape Town a member of BP South Africa's staff examined the computer print-out of the company's sales. For the months until mid-1977 there were eleven columns of figures showing BP deliveries to depots in the Natref Area. In August 1977 the number of columns had been reduced from eleven to ten; it was this that had led the Chairman of BP South Africa to claim that the swap had been ended. The BP staff member then examined the remaining ten columns for the period since mid-1977 more closely, and found that the later figures were slightly higher than would have been expected. Further inquiries revealed that the supplies which BP provided in compensation for the loss of its Freight Services trade had been disguised, by integrating them into the ordinary South African sales. The Sasol swap had therefore *not* been terminated – it had merely been hidden in the company's records.

This embarrassing discovery was made on 15 September 1978. Wayne Templer, the Chairman of BP South Africa, was told, and he immediately ordered that the Sasol swap arrangement should be terminated. The swap was in fact ended at midnight on the very same day, a clear sign that action to end this link with Rhodesian trade could in fact have been taken much earlier.

Templer now faced a painful dilemma. He realised that he ought to inform his head office in London, but hardly relished the prospect of telling Sir David Steel that incorrect information provided by him had resulted in the company issuing a misleading statement to the press. Templer was also worried that by informing BP over the telephone he might contravene the South African Official Secrets Act.

On 20 September, the day *after* BP's denial of the Sasol swap, Templer finally phoned London with the embarrassing news. Sir David Steel, who was furious, immediately ordered a further investigation. Geoffrey Butcher was yet again sent out to South Africa, this time accompanied by a Director of BP's main Board (Peter Walters) a company lawyer, and a QC. After a nine-day investigation, the team confirmed that the Sasol swap had indeed at last been ended. Sir David Steel, no doubt concerned at the way in which embarrassing information on Rhodesia had been leaking from his company, decided that BP had no choice but to go back to the *Sunday Times* to confess that his earlier denials had been misleading.

On 22 October 1978 the *Sunday Times* carried a BP statement which explained that although the company's response of 19 September had been strictly correct (since the swap had actually been ended four days earlier), its denial 'could have given a misleading impression of the situation' that existed at the time of the publication of the *Sunday Times* report.

Security at BP's Britannic House was once again stepped up. All files relating to trade with Sasol were collected from different offices and stored at a central registry. On 29 December 1978 a two-page memorandum was sent to all senior employees on BP's new Rhodesian policy. Any staff who came across an order which they suspected might be destined for Rhodesia were asked immediately to telephone the Regional Co-ordinator for Africa with the news. Particular attention, the memorandum added, should be given to 'any new or unusual business' with BP South Africa or Sasol. At last BP was trying to take effective action. It was only unfortunate that this memorandum had not been circulated thirteen years before, when sanctions were first introduced.

The other oil companies which had participated in the

Sasol swap – Shell, Mobil, Caltex, and Total – may possibly have followed BP's example and ended the *formal* arrangement over Rhodesian sales. Yet little actually changed in practice. Under the original 1971 agreement with Natref, signed when the refinery was established, the international oil companies still have to help fill any shortfalls in the Natref Area caused by Sasol's inability to meet demand.

This situation may easily have arisen early in 1979. The ending of Iranian oil sales to South Africa following the fall of the Shah severely affected the Natref refinery. The National Iranian Oil Corporation was contracted to supply the majority of its crude oil requirements; the cut-off of Iranian oil almost certainly led to a decline in Natref's output. In these circumstances the other oil companies in South Africa are obliged to meet any shortfall in the Natref Area, and by providing these supplies the oil companies are therefore enabling Sasol to continue exporting to Rhodesia.

* * * *

Even after the ending of the Sasol swap, however, it appears that Shell, at least, still retained some more direct involvement in supplying Rhodesia. A little-noticed revelation in the Bingham Report is the admission by Michael Pocock, the Chairman of Shell, that his company's lubricants are still being supplied to Rhodesia. He claimed that 'a very small proportion of the products manufactured from crude oil supplied to Shell Oil South Africa may reach Rhodesia in the form of base lubricating oils' (B13.3). This 'base-stock' is mixed with various additives at the Shell blending plant at Willowvale, near Salisbury, which had been built in 1974 with some of the profits accumulated in Rhodesia.

The Shell lubricant plant was built to serve the entire Rhodesian market. It apparently blends lubricants to Shell specifications, which are then put into tins marked with the trademarks of Mobil, Caltex, Total, and BP, as well as the famous Shell symbol. This means that the Shell plant imports virtually the entire lubricant requirements of the Rhodesian market. Pocock claimed that the 'base-stock' exported from South Africa to Rhodesia represents 'a very small proportion' of South African imports. But the Rhodesian supplies of 'base-stock' blended at the Willowvale plant probably represent a very large proportion of Rhodesia's requirements.

Shell also appears to be providing much of the crude oil which Sasol subsequently refines and sells to Rhodesia. Shell International Petroleum, a company registered in the UK, may now be supplying about a third of the Natref refinery's requirements. In 1977 the National Iranian Oil Corporation sold 35,000 barrels a day to Natref. Since the remainder of Natref's requirements were provided by Shell (according to the Bingham Report) and the refinery uses about 65,000 barrels a day, Shell shipments may have amounted to up to 30,000 barrels a day.

The crude oil supplied by Shell to Natref comes from the Seria oilfield in Brunei, on the north coast of the East Indian island of Borneo. Brunei's official trade statistics show that the country exported 24,000 barrels a day of crude oil to South Africa in 1976, a figure in line with the estimate given above for 1977. Brunei has an unusual political status: as a British-protected state, its foreign policy comes under the control of the UK Government.

In October 1977 Shell received an assurance from Sasol, the main shareholders in the Natref refinery, that 'no products produced from the crude oil which you will be supplying to us under this contract will be exported to Rhodesia' (B13.32). This assurance is puzzling. Until very recently Iran was the only other source of crude oil for Natref. Yet in 1978 the Shah's Foreign Minister claimed that he had obtained a written assurance from Pretoria that 'no drop' of Iranian oil was being delivered to Rhodesia.[12]

The Shell and Iranian assurances, it would seem, cannot both be correct. Two explanations have been suggested for the discrepancy. First, that Sasol's small oil-from-coal plant might be supplying Rhodesia. But this produces 4,500 barrels a day, which is only a quarter of Rhodesia's requirements. Secondly, Sasol might be refining oil from South Africa's large strategic stockpile. But again this seems very unlikely, due to the enormous logistical problems involved in pumping up stocks from the disused coal mines where the oil is stored.

It should be noted that 'assurances' which the British oil companies have received from South Africa regarding

Rhodesian trade have proved notoriously unreliable in the past. Certainly the British Government has remained somewhat sceptical of Sasol's claim that Brunei crude oil is not reaching Rhodesia in refined form. Foreign Office Minister of State Ted Rowlands admitted in April 1979 that the Government was simply 'not in a position to vouch for Sasol's observances of these assurances'.[13]

It is likely that oil products processed from both Brunei and Iranian crude oil have been reaching Rhodesia. In the refining process it is technically extremely difficult to separate different shipments of crude oil, and then to say that a particular barrel of refined oil originated from a certain consignment of crude oil. Some of the Brunei oil supplied by Shell must therefore probably end up in Rhodesia after refining.

In November 1978, after it had been revealed that Shell and BP had been involved in a swap arrangement with Sasol to supply Rhodesia, Dr Owen told Parliament that he had 'placed Shell and BP formally on notice of the Government's strongly held view that no company in the Shell or BP group should be involved in the supply of oil to Rhodesia, whether direct, indirect or by participation in marketing arrangements related to the supply of oil by others to Rhodesia'.[14] But sales of Shell crude oil to Sasol, the source of most of Rhodesia's oil, appear incompatible with Dr Owen's instructions.

*   *   *   *

The mounting evidence of sanctions-busting by Western oil companies that emerged during 1977 led to a major diplomatic offensive to tighten the oil embargo. In the autumn of 1977 the Organisation of African Unity sent a seven-nation mission headed by the Zambian Foreign Minister to visit the major oil-exporting countries in order to emphasise the importance of the embargo against Rhodesia and South Africa; the United Nations Sanctions Committee held a series of meetings devoted to discussing ways of tightening the oil embargo; the UN General Assembly adopted a number of resolutions calling for an extension of oil sanctions to include South Africa; and African members of the UN also called for a Security Council decision to impose a mandatory oil embargo against South Africa, in reprisal against the country's support for the Smith regime.

The Commonwealth was the most important international forum in which detailed proposals for tightening the Rhodesian oil embargo were raised. At the Commonwealth summit conference in June 1977 sanctions-busting dominated the discussions on Southern Africa. The failure of the oil embargo, the communiqué recorded at the end of the meeting, was 'a crucial factor in the survival of the illegal regime' in Rhodesia.[15] A special Working Group on Sanctions, which was set up to study the situation, reported to the Commonwealth Committee on Southern Africa on 21 no oil would be supplied to Rhodesia, or the embargo should be extended to include South Africa itself. The October 1977.

South Africa, the Commonwealth decided, must be given a 'stark choice'. Either South Africa should guarantee that Commonwealth decision was unanimous. But the British Government expressed reservations, and insisted that the time was not yet 'ripe' for new action over Rhodesia. African diplomats, recalling that the embargo had been introduced nearly twelve years earlier, wondered if the time would *ever* be ripe in British eyes.

International discussion of action to tighten the oil embargo undoubtedly increased pressure on the Rhodesian regime. Smith himself admitted, in November 1978: 'When pressure was applied on South Africa which involved Rhodesia, Pretoria always passed that pressure on to me.'[16] Recent threats of an oil embargo were a prime example: 'These things worry the South African Government. It would be stupid of us to think otherwise.'

*   *   *   *

It has sometimes been argued that since Rhodesian sanctions have failed, an embargo against South Africa would be equally ineffective. The two cases, however, are very different. Rhodesia's survival depends on support from a friendly neighbour (namely South Africa); but South Africa has no local ally with access to oil. South Africa itself is therefore dangerously vulnerable to an oil embargo.[17]

In 1978 South Africa imported about 400,000 barrels a

day of crude oil, and around 15,000 barrels a day of refined products. No domestic deposits of oil have been discovered, despite many years of prospecting. A small oil-from-coal plant, known as Sasol I, produces only 1 per cent of the country's requirements. The strategic stockpiles, largely kept in disused coal mines, would only last about one and a half years at normal rates of consumption. A total cut-off of oil imports would therefore bring South Africa to the brink of collapse within two years.

Early in 1979 South Africa faced the most serious threat so far to its oil supply. Since the 1973 oil crisis the country had depended on Iran for over 90 per cent of its oil requirements. The shah's Government had 'considered oil as a commodity and not a political weapon', and had refused to follow the Arab members of OPEC in imposing an embargo against the apartheid regime.'[18] South Africa, however, was dangerously dependent on a single exporter of oil; and the country's survival would therefore be threatened if that source was ever cut off.

Recent political developments in Iran have indeed had a severe impact on South Africa. The growing wave of strikes against the Shah's regime led to a complete halt to Iranian oil exports on 26 December 1978, and after the Shah fled abroad, early in 1979, the new Government headed by Dr Shapour Bakhtiar announced that Iran would never resume oil sales to South Africa. This position was maintained when the Ayatollah Khomeini flew back from exile in Paris to inaugurate an Islamic republic.

Brunei is now the only country which openly sells crude oil to South Africa. But Shell supplies from Brunei provide only about 5 per cent of the country's requirements. South Africa is therefore having to purchase the bulk of its oil through brokers, on what is known as the Rotterdam 'spot market'. This is very expensive, and South Africa now has to pay a heavy premium for its oil. In 1979 South Africa's oil bill is likely to double to over £1,500 million.

African diplomats have therefore pointed out that South Africa's vulnerability offers a new opportunity to enforce the oil embargo against Rhodesia: unless South Africa immediately cuts off all oil supplies to Rhodesia, they propose extending the oil embargo to cover South Africa itself.

* * * *

> # Shell and BP Marketing Services (Pvt) Ltd.
> ## thank most sincerely all those people involved in fighting the recent fire at the Salisbury Depot, as well as those who offered their assistance and support.
>
> ## Meanwhile, we are in business as usual.

Fig. 7

A copy of a Shell and BP advertisement that appeared in the Rhodesian *Daily Mail*, 31 December 1978. The advertisement refers to the guerrilla attack on the Shell/BP fuel storage tanks at Salisbury.

The situation inside Rhodesia has steadily deteriorated. Smith opted for an internal settlement, by excluding the guerrilla fighters, and on 3 March 1978 the Transitional Government was formed. Three African leaders joined Smith in the new administration: Bishop Abel Muzorewa (UANC), the Reverend Ndabiningi Sithole (internal wing of ZANU), and Chief Jeremiah Chirau (Zimbabwe United

People's Organisation). The white minority, however, effectively remained in control, and the internal settlement was therefore widely condemned both inside Rhodesia and at the United Nations.

Joshua Nkomo's ZAPU and Robert Mugabe's external wing of ZANU, which had joined to form the Patriotic Front, rejected the internal settlement, and continued to escalate the guerrilla war. One of the most dramatic incidents in the war was an attack on the country's main oil storage depot at Salisbury – a bitter reminder of the failure of sanctions to lead to a peaceful solution to UDI. On 11 December 1978 three big explosions ripped through the Shell/BP tanks. The fire quickly spread to the Mobil, Caltex, and Total tanks. When the blaze was finally brought under control, six days later, it was discovered that 28 of the 35 storage tanks had been destroyed. Over half a million barrels of refined oil products – worth some £10 million – went up in smoke. Smith himself admitted that the attack had been 'a great disaster' and one of the biggest setbacks in the guerrilla war. There could hardly have been a more visible sign that the war was reaching the very heart of the country.

The Rhodesian election in April 1979 did little to provide a political solution for the strife-torn country. On 1 June 1979 Bishop Muzorewa, whose UANC swept the polls, was inaugurated as the first black Prime Minister of the country which became known as Zimbabwe-Rhodesia. Even the name of the new state was an indicator of the fragile compromise which had been achieved. Political and economic power still continues to be largely in white hands. Smith himself remains in the new Government as Minister without Portfolio.

The new Conservative Government which came to power in Britain after the election of 3 May 1979 seemed set on gradual recognition of the Rhodesian regime. Conservative policy was to lift sanctions, on the grounds that 'majority rule' had been achieved. But such a move is unlikely before the Commonwealth summit conference, which is due to take place in Zambia in August 1979. Black African states, and particularly oil-rich Nigeria, will no doubt threaten to take economic retaliation against Britain if Margaret Thatcher's Government decides to lift sanctions before an internationally acceptable settlement is reached in Rhodesia.

A solution to Rhodesia's political problems now seems even more remote than it had when Ian Smith declared UDI. The Reverend Sithole has refused to participate in the new Government, on the grounds that the April 1979 elections were improperly conducted. Much more important, however, is the fact that the Patriotic Front has continued to condemn the 'internal settlement' because it does not represent a genuine black Government. Merely bringing a few black faces into the Government, the Patriotic Front argues, does nothing to alter the fundamental balance of power under which the whites continue to remain in effective control. Zimbabwe-Rhodesia is a country torn apart by civil war: a bloody conflict for which the oil companies must bear a heavy share of the blame.

NOTES

1 *Sunday Times*, 31 October 1976.
2 *Guardian*, 22 February 1978.
3 UN Sanctions Committee, 9th annual report, p. 304.
4 BP Submission to the Bingham Inquiry, 27 September 1977, p. 95.
5 BP Submission to the Bingham Inquiry, 27 September 1977, p. 93.
6 *The Times*, 11 September 1978.
7 *Daily Mail*, 7 September 1978.
8 Hansard, Commons, 7 November 1978, col. 708.
9 BP statement of 19 September 1978.
10 Statement of 19 September 1978.
11 *Agence France Presse*, 14 February 1978.
12 *Marches Tropicaux*, 3 March 1978.
13 Letter from Ted Rowlands to Frank Hooley MP, 19 April 1979.
14 Hansard, Commons, 7 November 1978. cols. 708–9.
15 Commonwealth Communiqué, para. 16.
16 *The Times*, 29 November 1978.
17 For further details see *Oil Sanctions Against South Africa*, United Nations, June 1978.
18 Address to Fourth Committee of UN General Assembly, 8 December 1977.

13   The Scandal

Oilgate, the *Sunday Times* declared, is a 'scandal reaching to the very heart of Government',[1] providing an unusual insight into the interaction of oil company executives, civil servants, and politicians. Its implications go way beyond the question of Rhodesia and raise key questions about where power really lies in our system of government.

Dr Owen, who commissioned the official inquiry into sanctions-busting, has argued that 'the Government and the country must face up to the implications of the Bingham Report'.[2] It is disturbing, however, that the political establishment has so far frustrated an attempt to investigate the *Government's* handling of the oil embargo. The basic questions that need to be examined by a further investigation are clear:

* *Did big business dictate Britain's policy on Southern Africa?*
* *How did Shell and BP escape prosecution?*
* *Did civil servants mislead their Ministers?*
* *Why was the Prime Minister so ignorant?*
* *Was the truth kept from Parliament?*

Did big business dictate Britain's policy on Southern Africa?

There have been increasing signs over the past few years that Britain's policy towards the minority regimes of Southern Africa has been unduly influenced by a handful of multi-national companies with extensive interests in the White South. The story of the oil embargo shows the way in which the powerful international oil companies frustrated the attempt to crush the Smith regime by sanctions. This obviously had a profound effect on political developments in Rhodesia, and in Southern Africa as a whole.

But there is also disturbing evidence that the British-owned oil companies were positively sympathetic to the white regimes. It is obviously quite legitimate for Shell and BP to hold discussions with the South African and British Governments on matters relating specifically to the oil industry. But Shell had also actually been pressing for an end to the British *arms* embargo against South Africa – a policy originally introduced to show the UK's abhorrence of the apartheid system.

The minutes of Thomson's meeting with the oil companies of 21 February 1968 reveal that the Shell Managing Director responsible for Africa, Frank McFadzean, opened the discussion by stating that his company had become 'increasingly worried about the deteriorating relations between South Africa and the United Kingdom' (B/A 259). One of the major reasons for this 'deterioration', he claimed, was the fact that Britain had decided not to resume arms sales to the apartheid regime. Later in the meeting, according to the official record, McFadzean 'interjected that he feared that our policy on the supply of arms would have repercussions in other commercial fields' (B/A 258).

Shell's position on the arms embargo had been explained in more detail in a note prepared eight days earlier. It stated that the 'recent cancellation of the arms deal has lent considerable force' to South African misgivings about British policy (B/A 250). Another Shell paper recorded that the arms embargo had had an 'extremely unfortunate impact' on South African public opinion (B/A 251) – a comment which suggested that black opinion in South Africa was quite irrelevant.

Shell's concern about the arms embargo is particularly interesting, since the United Kingdom-South Africa Trade Association, a group of British companies dealing with South Africa, was at this time also pushing for a resumption of arms sales. McFadzean's remarks therefore seem to have been part of a wider strategy by British business interests to 'normalise' relations with the apartheid regime.

The meeting with Thomson was held just a few weeks after it had been discovered that Shell Mozambique had been busting sanctions. Yet at the meeting McFadzean complained that Shell's efforts to respect sanctions were likely to hurt Britain's relations with South Africa – a country which was openly sustaining a rebellion against the Crown. McFadzean 'expressed considerable concern at the damage being done to our own – and British in general – interests in South Africa by the efforts we have made to enforce compliance with UN/British sanctions law'. At the meeting, he recalled, 'the strategic as well as commercial importance of South Africa was stressed' (B/A 283).

McFadzean appears to have had close personal links with the South African authorities. In November 1966 he met Prime Minister Vorster for discussions. On 25 November 1966, on his return to London, he wrote to Sir Paul Gore-Booth, the Permanent Under-Secretary at the Foreign Office, to outline his views on British-South African relations. Again, at a further meeting with the South African Prime Minister in mid-1968, McFadzean 'tried to reassure Mr Vorster about the British Government's intentions' (B/A 270). McFadzean, in reporting on these discussions to Thomson, made his own views crystal clear: Vorster, as far as he was concerned, was 'a moderate'; referring apparently to the possibility of a black Government in Rhodesia, McFadzean pointed out that Vorster 'was not prepared to see anarchy on South Africa's northern borders'.

Unfortunately, little information is publicly available on McFadzean's discussions with the South African leader. But Brian Sedgemore MP told Parliament: 'I suspect when the public learn about some of the meetings which went on between [oil] industrialists and the Head of the South African Government, their eyes will pop open at the way in which we are governed,'[3] and added 'We are discussing something that stands somewhere between moral corruption and treason.'[4]

The international oil companies are among the largest Western investors in South Africa and Rhodesia. Because they are involved in such a strategically important sector of the economy, however, they are inevitably subjected to considerable control by the governments of the countries in which they operate. The oil companies, whether they like it or not, are forced to serve the interests of the established regimes, and cannot avoid playing a political role.

What is disturbing, however, is the extent to which the international oil companies have positively helped to maintain the *status quo* in Southern Africa. Even Wilson himself was forced to come to this conclusion after a careful study of the Bingham disclosures:

> Some of my Honourable Friends are deeply concerned about the power of multinational companies. I have not shared, and I do not share their anxiety about many of them, and never have. But the Bingham Report . . . might suggest that their case has been, not overstated, but understated, certainly in respect of Shell, arrogantly asserting power with scant regard for responsibility.[5]

How did Shell and BP escape prosecution?

One of the most surprising revelations to emerge from the sanctions scandal is that it was not until 1978 that the Director of Public Prosecutions even *considered* the question of whether the oil companies should be prosecuted for breaking British law. Since UDI thirty-four British firms have been found guilty of sanctions-busting, and have been fined sums ranging from £10 to £50,000.[6] In December 1977 the Beck Company was fined for selling £27,000 worth of petrol *pumps* to the rebel regime. Yet ironically, no British firm has been prosecuted for supplying the oil which was actually pumped through this equipment.

The UK Sanctions Order makes it illegal for British companies or citizens to 'do any act calculated to promote the supply or delivery of petroleum' to Rhodesia. Yet for much of the period since UDI this law has been contravened by Shell Mozambique (renamed BP and Shell Mozambique in June 1977), a company which is registered in London. From December 1966 to January 1968 Shell Mozambique supplied half of Rhodesia's oil. After the Total swap arrangement had introduced, in January 1968, the British company still handled some residual supplies for Rhodesia. Then from the end of the swap in December 1971 until the closure of the Mozambique-Rhodesia border in March 1976, Shell Mozambique again provided half of Rhodesia's oil requirements.

266

The head offices of the oil companies in London themselves believed that Shell Mozambique had contravened British law. A Shell note dated 22 January 1968 pointed out that 'Shell Mozambique is doing a thing prohibited' by the Sanctions Order (B/A 242); the following day Hugh Feetham of Shell Centre drafted a letter to Louis Walker which stated: 'The arrangements which you have entered into in South Africa and Mozambique are in all probability such as to place Shell Mozambique and its Directors in a position of liability to prosecution' (B/A 244).

On 19 February 1968 John Francis predicted that the Commonwealth Secretary would decide that 'Shell Mozambique had contravened the [Sanctions] Order' (B/A 254). Yet apparently, when Thomson met the oil executives just two days later, the question was not even raised. The Government had already decided *not* to prosecute the company for what was undoubtedly the most serious contravention of the Sanctions Order since UDI.

Thomson later used a number of arguments to justify the decision not to initiate legal action against the oil companies. Revelations that Shell Mozambique had defied the Sanctions Order would, he claimed, have given 'a great boost to the morale of the Rhodesians'; they would have provided 'ammunition' to Britain's critics at the United Nations; and they would also have created 'a crisis' for the South African subsidiaries of the British oil companies and in general would have done 'great economic damage' to the UK's investments in South Africa.[7]

These arguments now appear to have little validity. The Rhodesians' morale was sustained by the continued flow of petrol – and prosecution of the major company involved in this trade would have endangered the country's supply. Revelations that Shell Mozambique had been busting sanctions would undoubtedly have sparked off considerable criticism at the United Nations – but legal action against the company would at least have given the impression that Britain was still serious about enforcing sanctions. There is also little evidence to suggest that prosecution of a British-registered company operating in Mozambique would have led to retaliation by the South African Government.

Thomson also argued that the Government had 'to take the most effective action to stop British oil going into Rhodesia' and claimed that it would be easier to do this 'with the cooperation of the oil companies'.[8] Again this argument is difficult to understand. If legal proceedings had been initiated, then the British oil companies would surely have done all they could to end their involvement in handling oil for Rhodesia. The fact that prosecutions were not seriously considered undoubtedly gave the impression within Shell and BP that the British Government was not serious about sanctions – a feeling that probably accounts for the fact that Shell Mozambique was again brought back into the supply chain when the Total swap was ended.

The British Government therefore seems to have made a grave error of judgement in not prosecuting Shell Mozambique in 1968. But who took this crucial decision? Thomson has claimed that the question of legal action was considered by the Attorney General and the Ministry of Power, and that the decision not to prosecute was a matter of 'high Government policy'.[9] It has also been suggested that a paper submitted to the Cabinet's Overseas Policy and Defence Committee in March 1968 included a recommendation that no legal action should be taken against the British oil companies.

But Lord Elwyn-Jones, ex-Attorney General, has disputed this account. He has claimed that he 'has no recollection whatsoever that when he was Attorney General in 1968 he was asked to advise about a case to prosecute Shell for sanctions-busting'.[10] The fact that no representative of the Attorney General's office attended the 21 February 1968 meeting with the oil companies (or was even sent a copy of the minutes) also appears to back up Lord Elwyn-Jones account.

There is no evidence in the Bingham Report to suggest that the Ministry of Power ever seriously considered the question of prosecutions. Naturally the Ministry was more concerned with the technical side of the oil trade than with the legal, political, and diplomatic aspects of the Sanctions Order. The other Government department dealing with the oil embargo was the Foreign Office. Yet Michael Stewart, who took over as Foreign Secretary in March 1968, has claimed that 'the question of prosecuting Shell Mozambique was never raised with me'.[11]

Thomson's comment that the decision not to prosecute

was a matter of 'high Government policy' implies that it was either decided by the Prime Minister or discussed in Cabinet. But only an examination of the official records for 1968–69 will show who it was who actually approved the decision not to take legal action against the joint Shell/BP subsidiary in Mozambique.

Even after the introduction of the Total swap, however, Shell and BP may still have been contravening the Sanctions Order. The British-based suppliers of oil to South Africa (Shell International Petroleum and BP Trading) could well have been committing an offence by providing oil in the knowledge that a regular proportion of their shipments to South Africa (about 7 per cent) would be swapped with Total, which in turn was supplying the Rhodesian subsidiaries of Shell and BP.

Certainly the oil companies themselves were aware of this danger. John Francis of Shell Centre explained in a note dated 19 February 1968 that 'the "cleansing" of Shell Mozambique . . . has resulted in the transfer of activities to the South African companies, which involves their British suppliers in contravening the Order in Council'. He added that 'the strong possibilities are that Thomson . . . will form the view that . . . Shell and BP as suppliers to South Africa are now contravening the Orders' (B/A 254). Arthur Sandford of BP also admitted in a note dated the same day that 'our parent companies as suppliers of crude to Shell South Africa may be vulnerable to the Orders in Council' (B6.68).

It was therefore surprising, to say the least, that the Government condoned the Total swap. But at the meeting of 6 February 1969, when full details of the swap were given to the authorities, Thomson told the oil companies that their legal position was 'quite defensible' (B/A 270). James Bottomley, a senior Foreign Office official, added that 'the legal position was sound and could be defended'. Altogether, the Bingham Report suggests that the Government gave very little attention to the possibility that the Total swap could still involve a contravention of the Sanctions Order.

The Sasol swap involved similar legal difficulties. The British-based oil suppliers still knew that a regular proportion of the oil they shipped to South Africa would be swapped with Sasol. It is interesting, however, that in this case Dr Owen subsequently sent details of the Sasol swap to the DPP – which suggests that the authorities may now be at least considering a stricter interpretation of the Sanctions Order.

It is not only British companies which are subject to the Sanctions Order, but also British citizens. Directors and employees of the oil companies with British nationality are therefore covered by the sanctions law, both in Britain and abroad. The two London-based executives who have been most intimately involved in Rhodesian affairs since UDI have been John Francis (Shell) and Arthur Sandford (BP). The two men took up their posts in 1967, shortly after Shell Mozambique began handling Rhodesian supplies, and they both learned early in 1974 that the Total swap had been ended and that Shell Mozambique was again involved in supplying Rhodesia.

But undoubtedly the most important individual in the Rhodesian oil story has been Louis Walker, head of Shell South Africa and Shell Mozambique until 1973. Because he was born in South Africa, and lived there most of his life, it has been assumed that he is a South African citizen, and therefore not personally subject to the UK Sanctions Order. Cynical observers have commented that the Bingham Report had singled him out as a scapegoat because he was a foreigner and therefore not liable to prosecution.

Walker, however, is actually of *dual* British-South African nationality, and therefore *is* subject to the UK Sanctions Order. A study of the annual returns of Shell UK shows that in 1976 his nationality is given as South African, and in 1977 as British. Shell's Company Secretary subsequently confirmed that Walker does indeed have dual nationality. He could therefore well be liable to charges that he has extensively contravened British sanctions law.

As head of Shell South Africa, Walker was ultimately responsible for the supply of oil to Rhodesia – which was in opposition to the *stated* policies of his head office in London. Yet in 1977, surprising as it seems, he was actually appointed Director of Public Affairs for Shell UK and thus made responsible for dealing with the company's relations with the British Government. After the publication of the Bingham Report, which produced so much incriminating evidence against him, it might have been thought that Shell

270

would have removed him from the Board of Shell UK. But he still remains a Director.

Walker's continued rise within Shell illustrates one of the disturbing aspects of the sanctions saga: the way in which so many of the people involved, far from being demoted or even sacked for their actions, have actually been decorated and promoted to high office. John Francis is now based in New York as President of the Shell-owned Asiatic Petroleum. Dirk de Bruyne, who was made a Knight of the Order of the Netherlands Lion, is President of Royal Dutch Petroleum in The Hague. Hans Pohl is based in Hamburg as head of Shell Germany. Frank McFadzean, since knighted, became Chairman of British Airways. Louis Deny, a holder of the French National Order of Merit, is Deputy Director-General of the Compagnie Française des Pétroles (Total). Denys Milne replaced Alan Robertson as Chief Executive of BP Oil. Alan Gregory, formerly of the Ministry of Power, is now Chairman of BP Trading's Executive Committee. Lord Greenhill, who was given a peerage after he left the Foreign Office, remains a Government-appointed Director of BP.

At this stage the DPP is still studying the evidence contained in the Bingham Report, and no decision has yet been reached on whether or not to initiate prosecutions. Behind the scenes, however, it is being suggested that it would be wrong to prosecute the oil companies for actions which were condoned at the time by the Government. Thomson's 'fingerprints' on the arrangements have therefore assumed a new importance.

The decision not to prosecute the oil companies back in the 1960s was a severe blow to Britain's sanctions policy. Oil supplies to Rhodesia made by the Southern African subsidiaries of Shell and BP either directly or through swaps have probably totalled more than 25 million barrels – around half of the oil imported by the rebel regime since UDI. The implications of this go beyond the question of Rhodesian sanctions; Shell and BP, the two largest companies in Britain, seem to have been treated *as if they were above the law.*

Did civil servants mislead their Ministers?

Since UDI, the British Government seems to have had incredibly little idea about how oil was reaching Rhodesia. During the early months of 1966 officials underestimated the quantities of refined oil products being sent into Rhodesia by road from South Africa and by rail from Mozambique. Soon afterwards, Shell and BP started railing oil into Rhodesia on the South African Loop Route – a fact that does not seem to have been discovered until it was revealed in the Bingham Report over twelve years later. In December 1966 Shell and BP began using the Direct Route from Lourenço Marques to supply Rhodesia. Yet apparently it was not until early 1968 that the authorities became aware of the involvement of Shell Mozambique in this trade.

The Total swap arrangement was set up in 1968 to handle supplies for Rhodesia. But although the Total swap was ended in 1971, it appears that this was not known to the British Government until seven years later. After the closing of the Mozambique-Rhodesia border, in March 1976, most of Rhodesia's oil was supplied by Sasol under a new swap arrangement. Yet the existence of the Sasol swap does not seem to have been known in Whitehall until the story was broken in the *Sunday Times*.

The Government's efforts to monitor the oil embargo could hardly have been less effective – even though the oil embargo was the most important form of pressure that could be exerted on the rebel regime. A handful of journalists and other investigators at work in 1976–78 managed to discover a great deal about how the oil embargo was being breached; yet incredibly, the Government, with its massive resources, uncovered virtually nothing.

Dr Owen appears to have realised that something had gone badly wrong. Seven years ago, in writing about the oil embargo, he argued that 'the quality of the advice from the Ministry of Defence, the Commonwealth Relations Office and the intelligence community needs to be assessed.'[13]

There were in fact four major sources of information which should have provided a steady flow of data. First, Britain has a corps of diplomats in Southern Africa. There were Consulates in both Lourenço Marques and Beira, an

Embassy in Pretoria, and until 1968 a 'Residual Mission' in Salisbury. Reporting on sanctions-busting was clearly within the responsibility of all these diplomatic posts.

Secondly, the British Government has a network of intelligence agents in Southern Africa. It is obviously very difficult to know exactly what information was being sent back from MI6 agents out in Southern Africa, but there is disturbing evidence to suggest that they proved extremely inept at ferreting out the facts. A recent study of the secret service has argued that 'the success of UDI in Rhodesia was due in part to MI6 misreading the situation there.'[13] It added that MI6's 'intelligence on Rhodesia was so bad for so long that George Wigg, when Paymaster-General, had to organise a private team to get the right answers.' Arthur Bottomley, who was Commonwealth Secretary until August 1966, has claimed that he 'doubted whether the intelligence services were monitoring the embargo'.[14] If Bottomley is correct, and oil sanctions were not being monitored during the first half of 1966, then it is likely that even less attention was devoted to discovering who was supplying oil to the rebel regime in later years. The apparent failure of MI6 to keep a close watch on the oil embargo raises important issues concerning the operation of Britain's intelligence service.

The Portuguese Government was a third source of information. Dr Salazar was anxious to show that it was the international oil companies, and not Portugal, which were responsible for the failure of sanctions. On at least three occasions (May 1967, March 1968, and January 1969) the Portuguese Government produced detailed reports giving a statistical breakdown of the numbers of railway tank-cars despatched to Rhodesia by each of the oil companies. But unfortunately the British Government consistently discounted information received from Portugal, on the grounds that it was 'unreliable', and seems to have made little effort either to investigate these reports or to have sought further material from Lisbon.

Finally, the oil companies themselves were an important source of information. There were obvious dangers in relying too heavily on the head offices of Shell and BP, since it was unrealistic to expect them necessarily to volunteer incriminating evidence. During the first three years of the oil embargo there were extensive contacts between the oil companies and the Government. But from February 1969 until June 1976 – a period of more than seven years – there were no discussions between the Government and the oil companies over Rhodesian sanctions. This suggests that little if any effort was made during this period to monitor the oil embargo.

The civil servants thus had a range of possible sources of information which, if utilised properly, should have enabled the Government to discover how oil was reaching Rhodesia. But the signs are that the officials responsible for watching over the oil embargo were remarkably lax in their efforts. They consistently seem to have avoided asking the rather obvious questions that might have led to the discovery that British companies were involved in Rhodesian trade; and when evidence to this effect did emerge, it was rarely followed up with an urgent and thorough investigation. Indeed, it is unusual to find evidence of any civil servants having taken the initiative to discover the facts about the oil embargo. All along, the civil servants, particularly those at the Ministry of Power, tended to reflect the views of the oil companies.

There have also been allegations that the civil servants involved may have actually withheld vital information from the politicians. Dr Jeremy Bray, MP, in speaking of his experiences as Parliamentary Secretary at the Ministry of Power, accused Angus Beckett, the head of the Ministry's Petroleum Department, of having 'deliberately blocked information going to Ministers'.[15]

On at least two occasions key civil servants were given information on a 'personal' basis, which suggests that they had been requested not to pass on the facts to their Ministers. In January 1967 Barry Powell, one of Beckett's assistants, received detailed statistical data on supplies handled at Lourenço Marques. Dirk de Bruyne wrote to Powell: 'I feel I can do no better than send this information on a personal basis' (B6.18). This suggests that this vital data may not have been passed on to Richard Marsh, the Minister of Power.

At a meeting at the Ministry of Power on 15 May 1968 Alan Gregory, another of Beckett's assistants, was given details of the Total swap from John Francis of Shell. At the end of the discussion Gregory asked whether he could pass on the information he had received to Robert Clinton-Thomas at the Commonwealth Office 'on a purely personal basis', so that 'if there were any further signs of Ministers wishing to sound off on this subject, the appropriate discouraging noises could be made' (B/A 264). This again suggests that news of the Total swap was perhaps not passed on to the Commonwealth Secretary.

The Petroleum Department of the Ministry of Power was the main Government Department dealing with the oil embargo during the 1960s. Yet it appears that the politicians who headed the Ministry were given virtually no information about the activities of their civil servants. Neither Richard Marsh (Minister 1966–68) nor Dr Jeremy Bray (Parliamentary Secretary 1966–67) are mentioned in the Bingham Report, and they appear to have played no part in the sanctions story. Indeed, Bray himself has admitted that until he read the Bingham Report he was 'not aware that the Ministry of Power was the primary channel of communication with the oil companies over the Rhodesian embargo'.[16]

The civil servants' lack of vigour in discovering the facts is illustrated by the way in which the Government gradually learned about the Total swap. By the beginning of February 1968 the head offices of Shell and BP knew how the swap arrangement operated, and at the meeting with Thomson on 21 February 1968, the Commonwealth Secretary was – in BP's own words – told 'something of what had happened'.[17] But the oil companies' hints about the swap were apparently not picked up by any of the six civil servants present.

The oil companies obviously considered carefully how best to break the news, and decided that the best occasion would be a meeting with Alan Gregory at the Ministry of Power, which eventually took place on 15 May 1968. (Just over a year later Gregory left the Government to accept a senior post with BP.) Gregory made little attempt to seek further information on how the swap operated, recognising that neither the oil companies nor the Government would 'wish to be too involved with the details' (B/A 264). According to Shell's record of the meeting, the question of whether the new swap arrangement might contravene the Sanctions Order was never even raised.

The oil companies consistently outmanoeuvred the civil servants by breaking embarrassing news in such a way that the full impact of the disclosures was only gradually revealed. Once officials had been given some part of the story, and no action had been taken, it became more difficult to act when the full facts were known. The oil companies successfully used this technique to stamp the Government's 'fingerprints' on the procedures used for supplying their Rhodesian subsidiaries. But were the civil servants genuinely unaware of the oil companies' manipulative methods – or did they realise how they were being used?

Why was the Prime Minister so ignorant?

One of the most bizarre aspects of the sanctions story is the apparent ignorance of Wilson about how the oil embargo was being broken. Right from the early days – when he predicted Smith's downfall in 'weeks rather than months' – the Prime Minister seems to have been ignorant of even the most basic facts about how Rhodesia was importing its oil. Indeed, his public statements right up to the 1970s suggest that he thought the oil was still being transported on the South African Loop Route, whereas from the beginning of 1967 it had all been railed directly from Lourenço Marques. It is obviously particularly disturbing that the Prime Minister himself should have known so little about how Smith was defying sanctions.

There appears to be a large measure of self-delusion in Wilson's optimistic early predictions of Smith's downfall. Lord Walston, Parliamentary Under-Secretary at the Foreign Office at the time of UDI, has claimed that the Government was 'too relaxed' and 'listened to the most optimistic forecasts instead of the more realistic one'.[18] Dr Owen has also pointed out that 'It is part of the whole tragedy of the handling of the Rhodesian rebellion that initially gesture politics . . . was seen as a substitute for hard-headed realism'.[19]

It is difficult to explain how Wilson can have remained so ignorant about the oil embargo, particularly since he tended to regard Rhodesia as his own special problem. Throughout his period of office, Rhodesia was Wilson's constant pre-occupation: as George Brown explained, 'It was very much a Number Ten spectacular from the beginning.'[20] It was the major issue that he faced at every Commonwealth summit conference and he faced it again during the series of attempts to reach a settlement on Royal Navy battleships. Yet Wilson signally failed to ask some very obvious questions about how Rhodesia was getting its oil.

Equally serious, however, is the way in which Wilson's Ministers and civil servants failed to correct his mistakes. Over the years Wilson made a number of bizarre and inaccurate statements about oil sanctions. Yet only rarely did his advisors appear to correct him privately. Perhaps the best example of this was the débâcle over the 'weeks not months' prediction. Virtually all the politicians and advisors dealing with the oil embargo were astounded by Wilson's statement to Commonwealth leaders. But it seems that little effort was made to raise the matter privately with the Prime Minister, in order to correct his misplaced optimism. It is disturbing that the Prime Minister could have been so out of touch with the realities of one of the most serious political crises that he faced in office.

Was the truth kept from Parliament?

At one crucial point during Thomson's meeting with the oil companies on 21 February 1968 the oil companies and the Government agreed on how to phrase answers to 'awkward' questions likely to be raised in Parliament. The Commonwealth Secretary, according to Shell's record of the meeting, pointed out that he 'would need to phrase carefully his reply' to questions in the House (B/A 261) and the Government's own account stated that the two sides agreed on the following formula to be used in Parliament: 'No British company is supplying POL [petrol, oil, and lubricants] to Rhodesia.' If MPs pressed on the question of whether oil products consigned by Shell Mozambique to the Transvaal were being diverted to Rhodesia, as indeed *had* been the case, the Government would reply: 'We have of course looked into the possibility,' and 'we are satisfied that this is not happening' (B/A 257). Finally, it was agreed that further 'details should be avoided so as not to give the South African Government any grounds for believing that the

British oil companies were interfering with supplies to South African customers'. In effect, this meant that information was to be withheld from MPs in Britain merely to avoid alienating a foreign Government – and one, furthermore which was actively sustaining a rebellion against the Crown. The record of Thomson's meeting of 21 February 1968 with Shell and BP provides a remarkably well-documented case of collusion between a Government Minister and private companies to formulate misleading answers to Parliamentary Questions.

Just two weeks later the Government was able to make use of these carefully formulated replies in Parliament. On 5 March 1968 Lord Brown, Minister of State at the Board of Trade, was asked by Lord Brockway about allegations that Shell and BP were supplying Rhodesia. He replied that 'investigations done into the activities of British oil companies leave Her Majesty's Government satisfied that the British oil companies themselves are not supplying oil to Rhodesia'.[21] The Government's statement was blatant disinformation: Parliament was given the clear impression that Shell and BP were not involved in sanctions-busting; yet only a few weeks earlier a British company, Shell Mozambique, had been supplying half of Rhodesia's oil; and even when the statement was made the same company was probably still providing some residual supplies of oil for Rhodesia. Lord Brown's reply, even *if* it had been completely true, was extremely misleading, since MPs were given the distinct impression that Britain's hands had always been clean over oil sanctions.

Just one month later an MP got very close to the truth. On 1 April 1968 Frank Judd asked how much of Rhodesia's oil came 'through Total and how much through other companies?' Goronwy Roberts, then a Minister of State at the Foreign Office, gave what had now become the standard Government reply. He preferred 'to say nothing which might reveal the extent of our knowledge of the source of the oil currently reaching Rhodesia'.[22] This presumably reflected the decision of the Commonwealth Secretary and the oil companies that 'details should be avoided'.

Parliament was also fed with massive doses of hypocrisy. On 27 March 1968, just one month after he had learned that Shell Mozambique had been handling oil for Rhodesia, the

278

Commonwealth Secretary told the Commons that there was one simple way that sanctions could be tightened: if other UN members supervised the activities of their firms 'as well as Britain does, the impact of sanctions would be substantially and significantly increased.'[23] This was just one of a long series of statements by British Ministers proclaiming that the UK's record over sanctions was second to none (see also pp. 84 and 203), the effect of which was undoubtedly to deter MPs from asking probing questions.

The most serious example of a completely untrue statement came in May 1977. This was soon after the Bingham Inquiry had been set up, when oil sanctions had already become an extremely sensitive issue; there is therefore little excuse for civil servants and politicians not giving the question the attention it deserved. Lord Brockway (who had asked a similar question in the House of Lords back in 1968) inquired about allegations that Shell and BP oil was *still* reaching Rhodesia. Goronwy Roberts, who was again a Foreign Office Minister of State, told Parliament in reply: 'Assurances of a substantial character from the highest level have been given us that no oil from British companies, including BP, finds its way directly or indirectly to Rhodesia.'[24] Yet only two months earlier BP had specifically admitted to Sir Anthony Duff, Deputy Under-Secretary at the Foreign Office, that BP *was still* supplying certain oil products to Rhodesia through Freight Services.

It is an unfortunate fact that until recently very few Parliamentary Questions were asked about the oil embargo. MPs seem to have neglected to inquire why the principal weapon of the Government's Rhodesian policy was failing to have any impact. But on the few occasions on which MPs did press them on the question of Rhodesia's oil supplies, Government Ministers were guilty of giving misleading, and sometimes inaccurate replies. Their constant suggestion that foreigners were to blame for the failure of sanctions also helped to allay MPs suspicions and deterred them from asking awkward questions about the possible involvement of British companies.

During the Parliamentary debate on the Bingham Report, in November 1978, former Foreign Secretary Michael Stewart was asked why MPs had not been told about the Total swap. He replied: 'It would not have been a wise thing to do.'[25] Stewart added that 'all who have held Government responsibility have had to tackle from time to time the problem of exactly how much of everything that they . . . ought to make public'.

Rhodesia is an unusual foreign policy issue, in that Parliament plays a particularly direct role in it: every November the Sanctions Order comes up for renewal. Yet, how was it possible for MPs to make a considered judgment on whether the Sanctions Order should be renewed, when the facts about the most important form of sanctions – the oil embargo – were deliberately withheld by the Government? The failure of the Government to lay the facts before Parliament suggests that MPs were treated as voting fodder, who would pass through the lobbies as directed by the whips. The oil sanctions saga is therefore a serious blot on Britain's democratic tradition.

*   *   *   *

A fictional CIA agent in a recent spy novel suggests that there are two views of history: the 'cock-up' and 'conspiracy' theories. The Oilgate scandal, with its devastating mixture of incompetence and deviousness, appears to add credence to both theories.

There is no neat explanation for Oilgate; no individual oil executive, civil servant, or politician can be singled out as the villain of the story. Oilgate is more disturbing. It is an indictment of a political system that allows a major aspect of the most important foreign policy issue which Britain has faced in recent years to be treated as a mockery – behind closed doors. But is Oilgate a story in which everything went wrong? Or is it just a more dramatic and well-documented example of the political process in Britain today?

NOTES

1 *Sunday Times*, 3 September 1978.
2 *Financial Times*, 5 October 1978.
3 Hansard, Commons, 1 February 1979, col. 1743.
4 Hansard, Commons, 1 February 1979, col. 1744.
5 Hansard, Commons, 7 November 1978, col. 754.
6 *Guardian*, 8 November 1978.
7 Thomson's statement of 6 September 1978.
8 Granada Television, *World in Action*, 9 August 1978.

9 Granada Television, *World in Action*, 9 August 1978.
10 *Sunday Times*, 3 September 1978.
11 Interview, 10 April 1979.
12 Dr David Owen, *The Politics of Defence*, p. 116.
13 Extract from Richard Deacon, *The British Connection*, (Hamish Hamilton, 1979), quoted in the *Guardian*, 29 May 1979.
14 Interview, 28 March 1979.
15 Hansard, Commons, 7 November 1978, col. 794.
16 Interview, 12 February 1979.
17 BP Submission to the Bingham Inquiry, 27 September 1977, p. 72.
18 *Guardian*, 20 September 1978.
19 Dr David Owen, *The Politics of Defence*, p. 114.
20 Hansard, Lords, 9 November 1978, col. 459.
21 Hansard, Lords, 5 March 1968, col. 1220.
22 Hansard, Commons, 1 April 1968, col. 30.
23 Hansard, Commons, 27 March 1968, col. 1668.
24 Hansard, Lords, 2 May 1977, col. 877.
25 Hansard, Commons, 7 November 1978, col. 823.

CAST OF CHARACTERS

The Cast of Characters lists the main individuals who have been involved in the sanctions story. The people have been divided into three main categories: oil company executives, British civil servants, and British politicians. Each entry gives the most important posts held by them at the time they participated in the story. The entry ends with details of the pages in which they appear in the book.

Oil Company Executives

Shell

BARREN, Sir David: Deputy Chairman of Shell Transport and Trading 1964–67. Chairman 1967–72. Now Managing Director.

BARRIE, Robert: General Manager of Shell/BP Rhodesia until 1972. Managing Director of Shell South Africa 1972–. 179, 222.

BÉNARD, A.: Managing Director responsible for Africa 1976–.

BERKIN, John: Managing Director responsible for Africa 1965–66. 134.

BONDS, E. J.: Executive dealing with Africa. 244–5, 251.

CHEW, Peter: Executive dealing with Africa. 228, 232.

DE BRUYNE, Dirk: Regional Co-ordinator for Africa 1965–68. Managing Director responsible for Africa 1972–75. Director of Shell Mozambique 1965–68. Now President of Royal Dutch Petroleum. 47, 82, 99, 111, 112, 113, 133, 134, 135, 176, 177, 178, 181, 187, 189, 203, 228, 231, 271, 274.

DE GEUS, A. P.: Regional Co-ordinator for Africa 1978–. 47, 99.

BRIDGEMAN, Sir Maurice: Chairman of BP 1960–69.

BIRKS, Dr J.: BP Trading Director responsible for Africa 1975–77.

BROOK, Sir Robin: Government-appointed Director of BP. 1970–73. 238.

BUTCHER, Geoffrey: Director of BP Trading responsible for Africa 1977–78. 248, 252, 254.

COLLINS, John: Head of Public Relations. 70, 251.

DRAKE, Sir Eric: Chairman of BP 1969–75. 228, 239.

DUMMETT, Robert: Managing Director responsible for Africa 1968–72. Director of Shell Mozambique 1965–67. Deputy Chairman of BP 1967–72. Now deceased. 85, 209, 215.

FRASER, William (later Lord Strathalmond): Managing Director responsible for Africa 1963–68. Director of Shell Mozambique 1967–69. Now deceased. 85, 121, 187, 190, 193, 194.

GOSS, R. M.: Executive in Public Affairs department. 52.

GREENHILL, Lord: Government-appointed Director of BP 1973–77. Formerly Under-Secretary of the Foreign Office. 52, 53, 91, 213, 217, 238, 271.

HARMER, Sir Frederic: Government-appointed Director of BP 1953–70. 191.

JACKSON, Tom: Government-appointed Director of BP 1975–. 52, 53, 68, 84.

JONES, Glynne: Regional Co-ordinator for Africa 1977–.

LAIDLAW, Christopher: Managing Director responsible for Africa 1972–. 46, 227, 232.

LOW, Cecil: Financial Director of BP South Africa 1974. 223, 240.

MCCUTCHEON, Ian: Managing Director of Servico, joint Shell/BP service company in South Africa 1973–. Director of Shell Mozambique 1973–75. 84, 227, 230, 231, 232, 252.

MILNE, Denys: Chairman BP South Africa 1971–75. Now Chief Executive of BP Oil. 41, 223, 225, 228, 229, 231, 232, 240, 251, 271.

MORRIS, Q. M.: Director of BP Trading responsible for Africa from 1978–.

PENNELL, Montague: Deputy Chairman of BP 1972–. 50, 249.

RIDDELL-WEBSTER, John: Regional Co-ordinator 1965–66. Director of Shell Mozambique 1964–66.

ROBERTSON, Alan: Director of BP Trading responsible for Africa 1971–75. 41, 225, 227, 228, 229, 231, 232, 271.

ROUNCE, John: Executive dealing with Africa. 83, 228, 229, 232.

SANDFORD, Arthur: Regional Co-ordinator for Africa 1967–75. Director of BP Rhodesia 1963–67 and 1971–75. Director of Shell Mozambique 1967–75. 79, 80, 82, 84, 85, 86, 87, 89, 92, 97, 188, 189, 196, 210, 215, 223, 227, 228, 229, 230, 232, 233, 251, 269, 270.

SAVAGE, Michael: Regional Co-ordinator for Africa 1975–77. Director of Shell Mozambique 1975–77. Now President of BP Alaska. 46, 47, 232.

STEEL, Sir David: Chairman of BP 1975–. 12, 90, 91, 252, 254.

TEMPLER, Wayne: Chairman of BP South Africa 1975–. Director of Shell Mozambique 1975–. 232, 252, 253, 254.

TRECHMAN, Gavin: Executive of Shell Mozambique 1973–75. General Manager Shell Mozambique 1975–77. Director of Shell Mozambique 1975–77. 232, 234, 235.

TREVELYAN, Lord: Government-appointed Director of BP 1965–75. 191.

WALLER, Kenneth: Regional Co-ordinator for Africa 1966–67.

WEBBER, Neil: Executive dealing with Africa. 225.

Caltex

CONLAN, Philip: Director of Caltex South Africa responsible for distribution. 102.

HARDY, Fred: Manager of Caltex's Mozambican operations.

MARSHALL-SMITH, William: Chairman of Caltex South Africa.

Mobil

ADAMS, Faneuil: Director of Mobil South Africa 1972–76. Now Vice President of Mobil Oil's International Division. 102.

BECK, William: Chairman of Mobil South Africa 1963–78. Chairman of Mobil Rhodesia. 32, 100, 102, 241, 243.

BIRRELL, George: Vice President and General Counsel of Mobil Oil. 32, 34, 37, 100.

CALVERT, John: Director of Mobil South Africa 1971. 102.

CHECKET, Everett: Director of Mobil South Africa 1971–72 and 1976–. Now Vice President of Mobil Oil. 102.

FOWLER, Edward: General Counsel of Mobil Oil's International Division. 37, 215.

GOMES, Fernando: Manager of Mobil's Mozambican operations. 33, 166.

GUBB, Nicholas: Executive of Mobil South Africa. 34, 52.

HALAMANDRIS, George: Former Manager of Mobil's Mozambican operations.

JACKSON, W. J.: Mobil Rhodesia executive. 52.

MASKEW, R. H.: Former executive of Mobil South Africa. 52, 205.

NICOL, J. Berwick: Managing Director of Mobil Rhodesia.

PHEAR, Stephen: Executive of Mobil Rhodesia. 37.

SCOTT, Charles: Director of Mobil South Africa 1968–71. 102.

SOLOMAN, Charles: Director of Mobil South Africa 1972–. 102.

VAN NIEKERK, Richard: Former executive of Mobil Rhodesia. 52, 205.

WILSON, D.: Mobil Rhodesia executive. 37.

Total

BANMEYER, Daniel: Director of Total South Africa 1958–. Chairman of Total South Africa 1975–. Chairman of Total Rhodesia. 103.

DALEMONT, Etienne: Director of Total South Africa. 103, 203.

DANNER, Duroc: Total director. 203.

DENY, Louis: General Manager of Total South Africa 1963–67. Now Deputy Director General of Compagnie Française des Pétroles. 103, 137, 166, 226, 271.

FLAMAND, Jean: General Manager of Total South Africa until 1972. 103, 137.

HOUGH, Alphonzo: General Manager of Total South Africa 1972–. Director of Total Rhodesia. 72.

KUMALO, Dumisani: Former executive of Total South Africa. 72, 73.

LISBOA, Eugenio: Former manager of Total's Mozambican operations. Now Cultural Attaché at Portuguese Embassy in London. 166.

MAROT, J.: Director of Information department. 103, 109.

PARQUET, A.: Director of Total South Africa. 103.

PARSONS, William: Head of marketing for Total South Africa. 72.

POMEROY, R. G.: Director of Total South Africa. 103.

Sonarep
BARTOLO, João: Former Manager of Sonarep. 240.

BOULLOSA, Manuel: Main owner of Sonarep. 140, 221.

GASH, Douglas: Former head of Rhodesia Oil Company. 240.

JARDIM, Jorge: Manager of Sonarep 1963–68. Now in private business. 5, 37, 60, 61, 67, 81, 86, 110, 111, 140, 141, 145, 150, 156, 158, 161, 165, 174, 175, 178, 193, 202, 218.

Castrol
THATCHER, Denis: Director 1967–76. 222.

WILSON, Stanley: President of Mobil East 1974. Now Managing Director of Burmah Oil.

British Politicians

AMERY, Julian: Minister of State at Foreign Office 1972–74. 219.

BENN, Anthony: Minister of Technology 1969–70. Energy Secretary 1975–79. 91.

BOTTOMLEY, Arthur: Commonwealth Secretary 1964–66. 116, 130, 139, 164, 273.

BOWDEN, Herbert (now Lord Aylestone): Commonwealth Secretary 1966–67. 172.

BRAY, Dr Jeremy: Parliamentary Under-Secretary at Ministry of Power 1966–67. 9, 181, 274, 275.

BROWN, Lord: Minister of State at Board of Trade 1968. 192, 278.

CALLAGHAN, James: Foreign Secretary 1974–76. Prime Minister 1976–79. Now Leader of the Opposition. 65, 66, 79, 92, 93, 98, 106, 235.

CARADON, Lord: UN Representative 1964–70. 152, 153, 154, 163, 204, 213.

CASTLE, Barbara: Employment Secretary 1968–70. 200.

CROSLAND, Anthony: Foreign Secretary 1976–77. Now deceased.

CROSSMAN, Richard: Social Services Secretary 1968–70. Now deceased. 119, 123, 145, 170, 193, 207, 215.

286

State at Foreign Office 1967. Commonwealth Secretary 1967–68. Now Chairman of Advertising Standards Authority. 9, 12, 17, 83, 92, 105, 106, 173, 182, 183, 184, 194, 195, 196, 198, 199, 200, 201, 203, 206, 208, 209, 210, 211, 212, 213, 215, 236, 237, 264, 268, 269, 271, 275, 277, 278, 280.

Civil Servants

Foreign and Commonwealth Office

LE QUESNE, Sir Martin: Head of Central African Department 1964–68. Deputy Under-Secretary 1971–74. Now High Commissioner in Nigeria. 198, 217.

LOMAS, Neville: Second Secretary at Embassy in South Africa 1961–66. Now at Foreign Office in London. 141.

MACKILLIGAN, David: Private Secretary to Commonwealth Secretary 1968–69. Now at Foreign Office in London. 198.

MELLOR, D. I.: Head of Rhodesia Economic Department 1977. 80, 97.

MONSON, Sir Leslie: High Commissioner in Zambia 1964–67. Deputy Under-Secretary at Commonwealth Relations Office/Foreign Office 1967–72. 212.

PALLISER, Sir Michael: Private Secretary to Prime Minister 1966–69. Minister at Embassy in France 1969. Under-Secretary 1975–. 45, 46, 198, 211, 212, 247, 248.

PUMPHREY, John: High Commissioner in Zambia 1967–71. 192.

RICHARD, Ivor: Representative to the UN 1974–79. 65, 66.

RENWICK, Robin: Head of Rhodesia Department 1978–.

ROSS, Sir Archibald: Ambassador in Portugal 1961–66. Now deceased. 174.

SNELLING, Sir Arthur: Deputy Under-Secretary 1962–70. Ambassador in South Africa 1970–72. 119.

STEPHENSON, Sir Hugh: Ambassador in South Africa 1965–66. Now deceased.

TAYLOR, John: Consul in Beira 1963–67. 160.

WHITE: former Consul in Lourenço Marques.

WRIGLEY, Charles: Consul at Beira 1969.

Ministry of Power

BECKETT, John: Head of Petroleum Division 1964–72. Now Chairman of William Press, North Sea contractors. 134, 181, 218, 274.

GREGORY, Alan: Assistant Secretary Petroleum Division 1968–70. Now Chairman of BP Trading. 205, 206, 210, 271, 274, 275.

MCNEIL, H. J.: Chief Executive Officer Petroleum Division 1966–68. 175, 186.

PITBLADO, Sir David: Permanent Secretary 1966–69. 134.

POWELL, Barry: Assistant Secretary Petroleum Division 1965–68. Now Director of Welsh Development Agency. 179, 181, 186, 191, 274.

Other Departments

JONES, Gruffydd: Private Secretary to Cabinet Secretary 1967–69. Now Under-Secretary at Welsh Office. 214.

NIELD, Sir William: Under-Secretary at Department of Economic Affairs 1964–66. 131.

TREND, Lord: Cabinet Secretary 1963–73.

WRIGHT, Sir Oliver: Private Secretary to Prime Minister 1964–66. Now Ambassador to Germany. 131.